That is why
theory of special relativity
is nonsense!

My motto:

When we study physical phenomena,
we always make
a mathematical model of them.

In such a model,
there are built-in physical laws
that are held together
by mathematical tools.

If the description of the physical phenomenon
is correct,
the mathematical model is
without contradictions
and
without paradoxes.

***

ex nihilo nihilum

Parmenides
500 BC

Jan Slowak

That is why theory of special relativity is nonsense!

Previous books

1. Bye-Bye Big Bang,
   Episod/Episode 1, 2, 3; swedish-english
4. **Redshift factor, Absolute redshift, Galaxies red / blue distribution**
5. Sawing of my book about the Big Bang
6. Big Bang –
   Questions to physicists and cosmologists
7. Einstein's special theory of relativity – mathematical and physical mistakes!
9. Back to Newton, swedish-english
10. Einstein's special theory of relativity = mathematical and physical nonsense... swedish
11. SR(LT, LF, TD, LK) = NONSENS, swedish
12. **Special Relativity is Nonsense, third edition**
13. **Light - The Absolute Reference in the Universe, fourth edition**

Publisher: BoD – Books on Demand, Stockholm, Sweden
Print: BoD – Books on Demand, Norderstedt, Germany
ISBN: 978-91-8027-735-8

For
Science

Content

1. Prologue ... 7
2. Space, Time, Event ... 12
3. A momentary picture of reality ... 24
4. LT - The most illogical concept in Physics ... 28
5. Alternative Lorentz transformations ... 57
6. LT are the culprit in the drama ... 75
7. Analysis of k-calculus ... 78
8. Experiment that support SR? ... 100
9. Horizontal light clock ... 105
10. Doppler effect/ Doppler shift ... 109
11. The pole in the barn paradox ... 114
12. Summa summarum ... 116
13. My last derivation of LT ... 121
14. A few excerpts from my notebooks ... 128
15. My correspondence with 'established researchers' ... 137
16. My certificate in SR ... 171
17. A small question for great researchers ... 172

1. Prologue

Why is it like that?
Why do some researchers, perhaps most of them, accept the special theory of relativity while a smaller number of researchers oppose it?
Everyone has read the same math and the same physics. Despite this, people think differently.
Do you think differently or do you just accept?

Much has been written about this. I do not intend to write the same thing again. Some links:

1) Jean de Climont: *The Worldwide List of Alternative Theories and Critics*
Here they have collected most of what is published and which deviates from "recognized" theories in the academic world.

2) Claes Johnson: *CJ on Mathematics and Science, blog*
Here, Claes Johnson presents his own theory, Many Minds Relativity, but what surprised me most was his blog posts, for example:
Special Relativity as Fake Physics
Why Einstein's Special Theory of Relativity is so Confusing
Einstein Did Not Understand the Idea of Relativity

Why so Difficult to Discuss Special Relativity with Physicists?
The Special Theory of Relativity = Pseudo-Physics?
Einstein: The Illusionist

Feel free to read from Claes Johnson's blog:
Dingle Destroyed as Scientist by Questioning Relativity Theory

How is it possible that a professor of mathematics writes like that? I'm talking about Claes Johnson. He is Professor of Applied Mathematics on the Royal Institute of Technology in Stockholm, Sweden. Not many people dare to do that.

How is it possible that one professor enthusiastically teaches the special theory of relativity, such as Brian Green, Columbia University, and another rejects it with the same enthusiasm and perseverance, such as Claes Johnson from the Royal Institute of Technology in Stockholm?

How is it possible that some who have worked and defended SR receive the Nobel Prize and others who reject it are ignored and ridiculed?

These professors, either are they followers of SR or they are deniers of SR, all these professors are paid primarily by tax money! How is it possible to teach SR all over the world but there are still established researchers who claim that this theory is wrong in one way or another?

One could perhaps accept this status quo in politics, but here we are talking about science, physics, logic and not least mathematics.
And in mathematics the claim that
"1 = 1" is true
while the claim that
"1 = 0" is false!

And what should I say about my research, I who do not even work at any university or college but hardly dare to call myself an independent researcher?

Could it be that it is primarily mathematicians who consider SR to be wrong?
Could it be that the mathematics is used deficiently in a way? It wouldn't surprise me.
In fact, a few years ago I heard a scientist who won a Nobel Prize complain that mathematicians are interfering with their work! It was on TV.
What? Why they use the math if they complain that the real mathematicians question their use of the math itself.

Mathematics is the queen of science!

Researchers in fields other than mathematics must not use mathematics in an incorrect way.
If you use the math, you have to do it the right way!
No sorcery may be used, no paradoxes may occur!

Despite this, I dare say that I have the best and most polished arguments that show that SR is nonsense, that SR does not verify reality.

Feel free to read about my previous arguments in my books:
B1: *Special Relativity is Nonsense*
B2: *Light - The Absolute Reference in the Universe*
and my articles:
A1: *Mathematics shows that the Lorentz transformations are not self-consistent*
A2: *Lorentz Transformations And Time Dilation Do Not Verify Reality*
A3: *Lorentz Transformations - The Sound versus The Light*

But just reading is not enough! One must discuss, question, involve as many as possible in this discussion.

In what follows, I present my new arguments against SR.

Even if it feels like I're repeating myself, that the pictures I present are similar, do not be influenced by this feeling.

2. Space, Time, Event

When talking about the special theory of relativity, SR, it is inevitable to talk about space, time and event. These are basic concepts that you have to deal with. The first two are so intricate that most physicists, even mathematicians and philosophers, all these thinkers of science, have processed them, pondered over them.

The conclusions from SR tell us that for reference systems in motion, space will be compressed, the concept of longitudinal contraction, that time will be extended, the concept of time dilation.
It is as if these two concepts, space and time, were material. But they are not! What particles does the room consist of? What particles do time consist of?

These conclusions from SR, length contraction and time dilation, seemed very strange, almost absurd, ever since my high school studies. What seemed completely unacceptable was the twin paradox.

Ask anyone what they mean by the space and you get as many answers as people you ask.

Ask anyone what is meant by time and you get as many answers as people you ask.

I do not engage in these discussions but I **argue that these concepts are not material and then they can not be contracted or dilated**.
But it's easy to say and harder to prove. In any case, I will try to prove that SR is nonsense. Because I have a lot of arguments that I have published in my books and articles. And in this book, I come up with a few more.

When we in SR talk about an event, it is referred to one or more inertial reference systems. Then we talk about reference systems that move in line with constant speed.
As everyone knows, we perceive our environment as a 3-dimensional reference system. A point in this reference system can be denoted by *3* coordinates *(x, y, z)*, length, width and height.

For the most part, we will work with only points on the x-axis and then our events will have only *1 (one)* spatial dimension.

When we talk about an event, we say that it occurs in space and time. We will denote an event with the letter E (event). We refer to a reference system with the letter S.

An event that occurs at the point x on the x-axis at

time t will be denoted by (x, t).

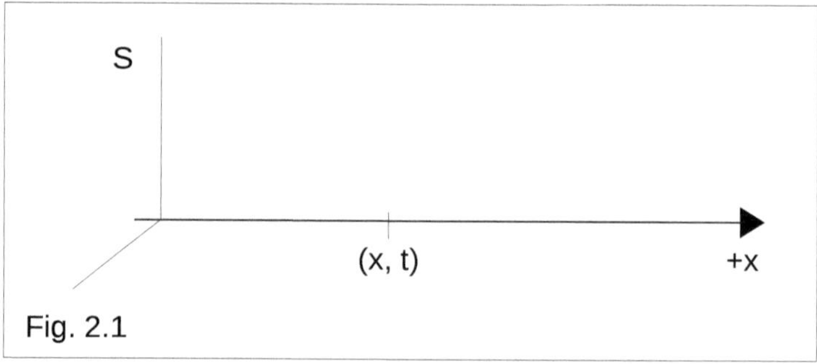

Fig. 2.1

An event that occurs in a reference system is not dependent in any way on this reference system at all. Only we choose to refer the event to S. There may be other reference systems to which we choose to refer the event. Say that there is a reference system S' located on the x-axis at a distance d from S.

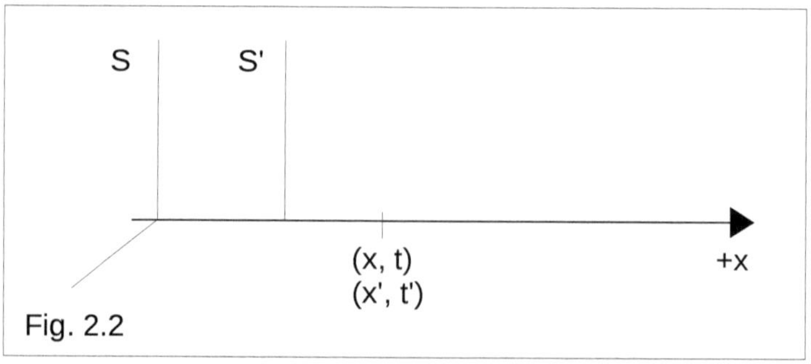

Fig. 2.2

Then the event will have coordinates (x, t) in S and (x', t') in S'.

But it does not matter if S or S' are there, if they are at

rest relative to each other or if they are in motion. They can not affect the event in any way. Nor can the event affect reference system S or S'. We can say that reference system S, reference system S' and event E are independent phenomena, objects.

We will now look at an event E in an inertial reference system S. In the beginning, we consider S as **a reference system in absolute rest in space**.

What we will do is define a method of how we determine the coordinates x and t. This is of utmost importance if we want to make comparisons between different events in a reference system or compare the coordinates of the event in two different reference systems.

We consider Fig. 2.1. When it comes to the coordinate x, it is quite easy for us to decide what it should be like. We determine that the origin of the reference system is at the point $x = 0$ and that the positive values are at the right. Then it is clear that the coordinate x is the distance between the S-origin (point $x = 0$) and the point on the x-axis where the event occurred.

When it comes to time, the time coordinates, it's a little harder. What moment should we take as $t = 0$, as the origin of the time axis. It's not so obvious.

It is always a convention we make.

Should we start counting the time in our reference system at lunchtime? When does the clock strike 12:00? But lunchtime is an event that has its time coordinate determined based on other criteria and agreement. In the case of Fig. 2.1, we could decide that we start counting the time just when the event occurs. We say the time was right $t = 0$ then. We do so in this case. We make a definition of the coordinates t. In other contexts, a different definition, a different time, a different agreement may fit.

We can also determine that it is the same time in all points in the reference system S. If the time in point x_1 is t then the time in point x_2 is also t!

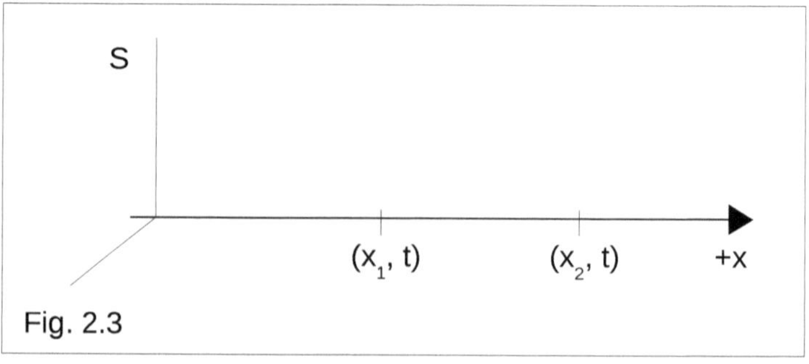

Fig. 2.3

I dare to go on to say the following:
If we can determine that all points in the reference system S have the same time then we can safely say that the time is the same in all other reference systems.
Another reference system S' is at a certain point x_0 in the reference system S and then S' can assume the same time as in S!
Time in x_0 is the same than time in point $x = 0$, the origin of referense system S!

What is stated in SR, that the time in a reference system depends on the speed of the reference system is just ridiculous! Why?
Imagine one point on the equator and one on the North Pole. The Earth's speed at the equator is about *1,670 km / h* (about *0.46 km / s*) but in this respect the speed at the pole is zero. It is believed that one second at the North Pole is not the same length as one second at the equator. But keep in mind that a point at the equator or the point at the North Pole has no time. The earth has been spinning around its axis for billions of years!

We return to Fig. 2.1 and the event $E(x, t)$. Say that the time is t when the event occurs at the point x. But how does S find out that an event has occurred at the point x? If the distance is small, an observer can see with his own eyes that something has happened at the point x. Seeing means that the light signal from the event, from the point x, reaches our eyes. We can say that the transmission of the information between the reference system and the event takes place by means of light signals.

What time is it in S when the light signal from E arrives? We denote this time by t_s.

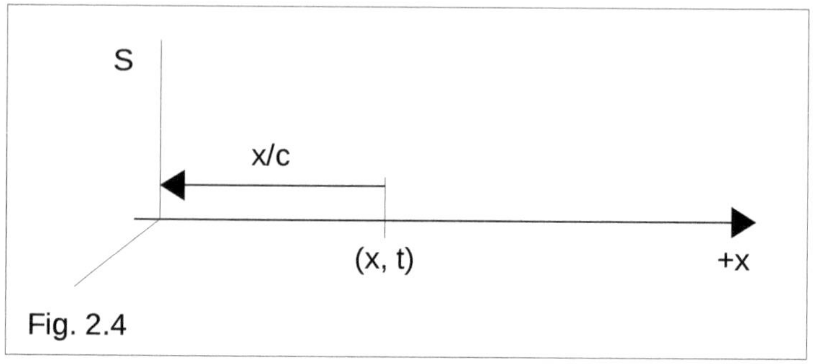

Fig. 2.4

The light signal needs the time x/c to go from E to S. Here, c is the speed of light, $c = 299{,}792.458\ km/s$, is approximated to $c = 300{,}000\ km/s$.
Then we can decide

$$t_s = t + x/c$$

and then S can calculate the time when the event occurred:

$$t = t_s - x / c$$

We see that we can calculate the time coordinate only if we know the distance to the event!
Here we have determined two different times, the time t when the event occurred and the time t_s when S received the information that the event took place.
And here comes a tricky question: which of these two times do we designate as the t-coordinate of the event E in the reference system S? What is more logical?
This is an important issue and is not addressed in the special theory of relativity. Why?
What happens if we have one or two reference systems and one or more events? It must be determined how the t-coordinate is defined.
When we decide on a method to define the t-coordinates then the same method should apply throughout the whole experiment.

We will perform many thought experiments. In many of them we will count the time from the time when the experiment begins.

For example:
The reference system S is at absolute rest in space and the reference system S' moves from left to right at a constant speed $v > 0$.
When S and S' are at the same point, their clocks are

synchronized, the time is set to zero, $t = 0, t' = 0$.

When time shows $t = 1$ in S, S' will be at the point $x = v$. The distance is calculated with the formula $x = vt$. When S' reaches the point $x = v$, its time t' shows and the distance to S is calculated, $x' = vt'$.

But we have one and the same distance.
This means that

$$vt = vt' \rightarrow t = t'$$

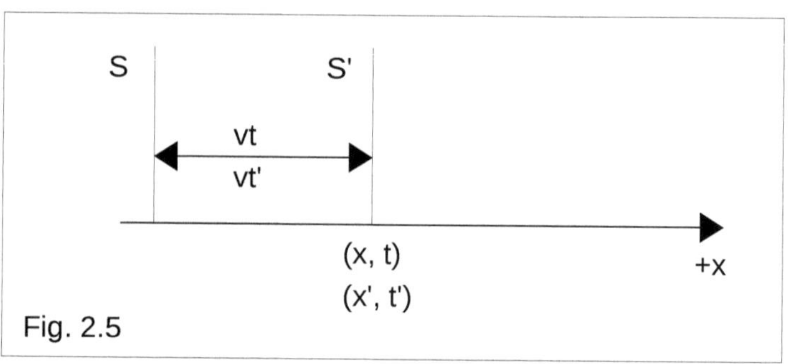

Fig. 2.5

Therefore, we conclude that the time is the same in all points and in all reference systems! There can be no time dilation!

We continue and look at some concrete cases. We consider two inertial reference systems, S in absolute rest in space and S' which is to the right of S at the distance d. An event occurs between these two reference systems.

Say that S has determined the value of the event coordinates to *(x, t)*. Can we express *(x′, t′)* using *(x, t)*?

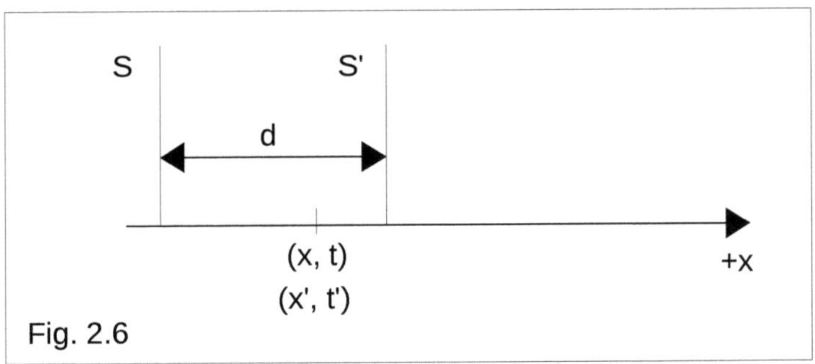

Fig. 2.6

$$x' = x - d$$
$$t' = t$$

We see that event is to the left of S' therefore x' will have a negative value. The time coordinate is the same, say it is the time when the event occurred.

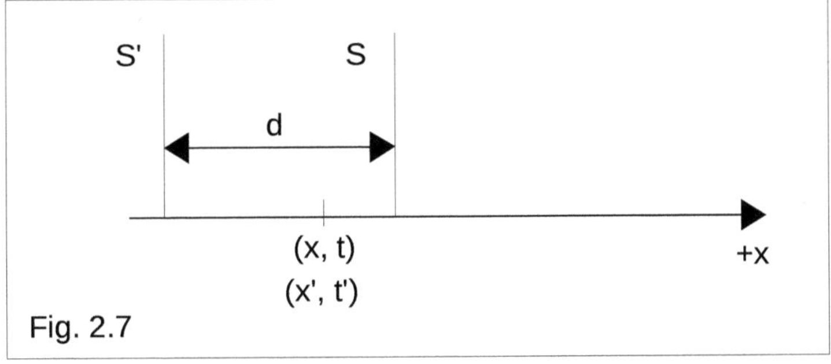

Fig. 2.7

In Fig. 2.7 we place S' to the left of S and calculate (x', t').

$$x' = d + x$$
$$t' = t$$

The coordinate x is negative, the event is to the left of S. The coordinate x' is positive, the event is to the right of S'.

We have two formulas for x' and we have not analyzed all the variants.
It is certainly possible to develop a general formula.

Now we will consider cases where S' moves to the right at a constant speed $v > 0$.
Fig. 2.8 is identical to Fig. 2.6 seen from the point of view of mathematics. The only difference is that instead of d we have vt or vt'!

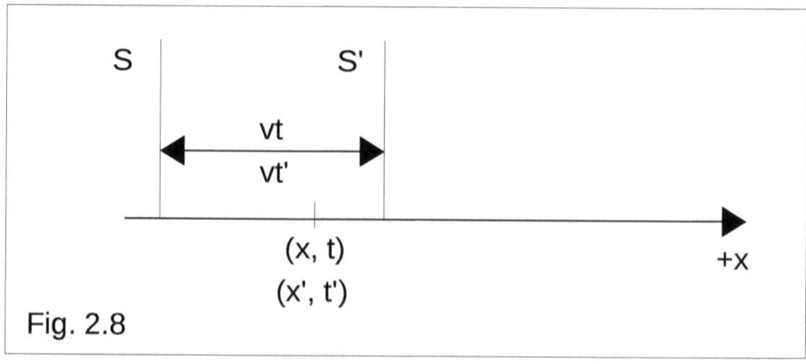

Fig. 2.8

$$x' = x - vt$$
$$t' = t$$

or

$$x = x' + vt'$$
$$t = t'$$

The most important insight here is the fact that the case when S' is in motion is from the point of view of mathematics identical to the case when S' is at rest relative to S.

The mathematical model, the mindset, the formulas are the same.

The mathematical model and formulas represent a momentary picture of reality.

3. A momentary picture of reality!

When we analyze a physical phenomenon from reality, we can make a mathematical model of it. The reality is complex, with many parameters and data. We can not include everything in the model and that is not the intention either. We take only the most necessary parameters and data.
For example, if we want to describe how a car moves on the highway, we do not need to draw the whole car with its four wheels, windshields, number of passengers.

We represent the highway as a line, say the x-axis. And we represent the car as a point. We determine a point that will be the origin of the reference system, S. A point x on the x-axis where an event E occurs, then represents the distance between points S and E.

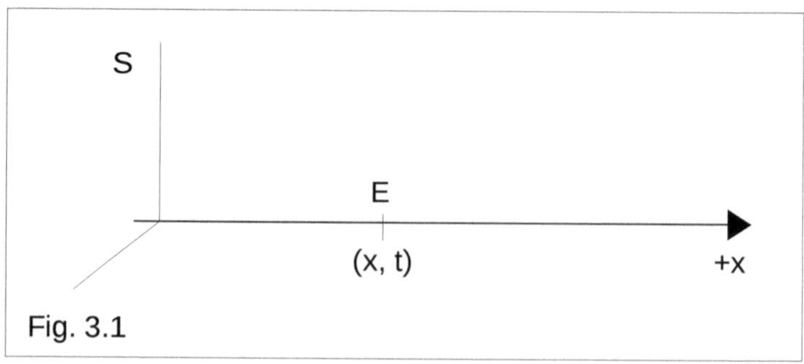

Fig. 3.1

Imagine that event E is the moment when a light

comes on at point E on the x-axis.
The reference system S finds out this when the light signal reaches the point S.
The light signal (the light), has many different parameters but we are only interested in the speed of light, c. Even when it comes to the car, it has different properties, it can run on petrol, diesel or electricity, but here we are only interested in its speed, v. And we will analyze, consider only the distances where the car moves at a constant speed. We will consider the car or another object as a reference system.

Say that the car S' moves from left to right on the x-axis at a constant speed v. When S' passes S, the clocks are set to zero. After time t, the car will be at the point $(x, t) = (vt, t)$.

But keep in mind that we actually have no clocks in the reference systems we analyze. All calculations we make regarding time, we do it from the point of view of mathematics. We deal with the mathematical time. We have no other option in terms of time and space in our thought experiments.
Do not forget this!

An event that occurs in a reference system is not dependent in any way on this reference system at all.

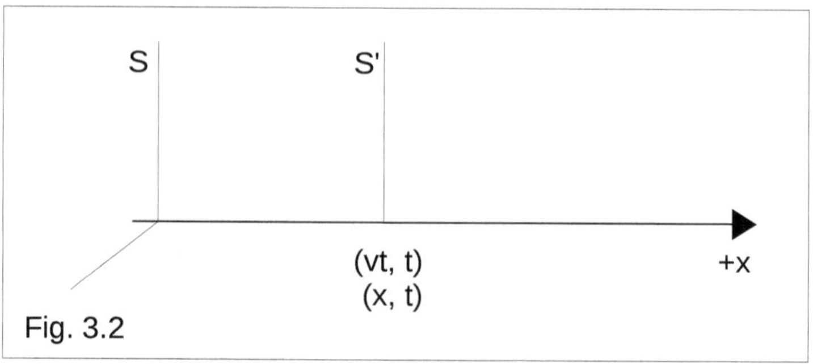

Fig. 3.2

Only we choose to refer the event to S. There may be other reference systems to which we choose to refer the event.

Reference systems S and S', event E, x-axis all these objects and phenomena are only mathematical elements in

the mathematical model

we create and which represent

a snapshot of reality.

A simplified picture of reality.

A reference system is only a point. In its movement, it does not take with it any space or time as if they were some kind of tails that have some properties, atributes. No. They are just mathematical concepts, mathematical elements in a system we create and can process with the help of mathematics and logic.

We create a mathematical model of reality. After we have created it and decided on a method, how to calculate the x-coordinate and the t-coordinate, then only mathematics applies.

In parts of the book that follows, I present various arguments and ideas from my research. All of these come to the conclusion that Lorentz's transformations apply only when S and S' are at rest relative to each other, when $v = 0$. Or when S and S' are at the same point. Then no theory of special relativity is needed to describe relationships between them.

4. Lorentz transformations – the most illogical concept in physics

4.1 Historical Facts

1864: James Clark Maxwell publishes the Dynamical Theory of the Electromagnetic Field. In this work, he derived that light is an electromagnetic wave. The light propagates in a medium in the same way that sound waves require air, and water waves require water. This medium is called an ether – light-bearing ether.

1887: The Michelson–Morley experiment is conducted. The purpose of the experiment was to confirm the existence of the ether and measure the Earth's velocity in space.

1904: Hendrik Lorentz formulates Lorentz transformations because the Michelson–Morley experiment could not demonstrate the existence of the ether.

1905: Albert Einstein publishes the special theory of relativity in which he included and used Lorentz transformations.

This work is divided into several sections in which we

analyze concepts from SR and ask questions about these concepts.

4.2. Mathematics, Cartesian coordinate systems

4.2.1 Mathematical definition for distance calculation

To calculate the distance between two points in a Cartesian coordinate system, the Euclidean distance is used.
This is done using the Pythagorean theorem; it is therefore also called the Pythagorean distance [4.1].

We consider a 3-dimensional (3-D) Cartesian coordinate system where we have the following two points:

$$A\ (x_1, y_1, z_1),$$
$$B\ (x_2, y_2, z_2).$$

The function for calculating the distance is denoted by *d*.
Then, we get the following:

$$[d(A, B)]^2 = (x_2 - x_1)^2 + (y_2 - y_1)^2 + (z_2 - z_1)^2.$$

In a 2-D Cartesian coordinate system, the distance becomes

$$[d(A, B)]^2 = (x_2 - x_1)^2 + (y_2 - y_1)^2.$$

Moreover, if we consider the 1-D Cartesian coordinate system (e.g. x-axis), the distance becomes

$$[d(A, B)]^2 = (x_2 - x_1)^2$$
$$d(A, B) = |x_2 - x_1|.$$

We always consider that A is to the left of B, that is, $x_1 < x_2$:

$$d(A, B) = x_2 - x_1.$$

This applies to all Cartesian coordinate systems!

4.2.2 Distance between two points in two different Cartesian coordinate systems

We consider two Cartesian coordinate systems, S and S'. S' is at a distance *a* from S. There are two points on their common x-axis:

for S, we have $A(x_1)$ and $B(x_2)$.
for S', we have $A(x'_1)$ and $B(x'_2)$.

According to Fig. 4.1, the following relationship exists between the x-coordinates of A and B:

$$x_1 = a + x'_1 \rightarrow x'_1 = x_1 - a,$$

$x_2 = a + x'_2 \rightarrow x'_2 = x_2 - a.$

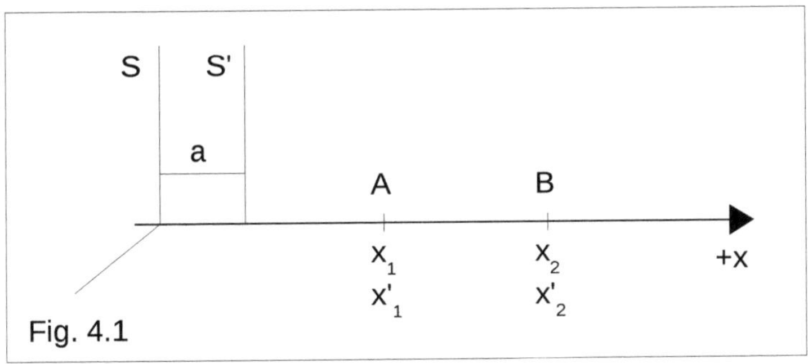

Fig. 4.1

We calculate the distance between points A and B:

for S: $d(A, B) = x_2 - x_1$
for S': $d(A, B) = x'_2 - x'_1 = (x_2 - a) - (x_1 - a) = x_2 - x_1.$

This means that **the Pythagorean distance gives the same distance between two points irrespective of the Cartesian coordinate system**.

4.2.3 Distance between two points in two different Cartesian coordinate systems that are in motion relative to each other

We consider two Cartesian coordinate systems, S and S'.

S' moves to the right relative to S at a constant speed, $v > 0$.

When the experiment begins, $t = 0$ and $t' = 0$, the two Cartesian coordinate systems are at the same point.

At time $t > 0$ and $t' > 0$, the two Cartesian coordinate systems are at distances vt (vt') from each other, Fig. 4.2.

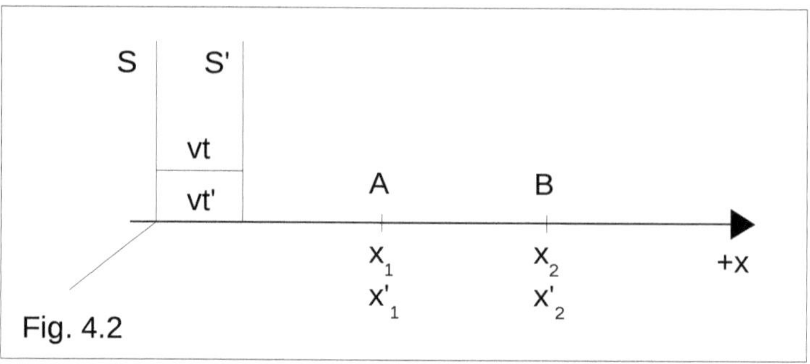

Fig. 4.2

The only difference between Fig. 4.1 and Fig. 4.2 is that the distance between the two cartesian coordinate systems is either vt or vt', depending on the system. We calculate the distance between points A and B:

for S: $d(A, B) = x_2 - x_1$
for S': $d(A, B) = x'_2 - x'_1 = (x_2 - vt') - (x_1 - vt') =$
$x_2 - x_1.$

Even in this case, the Pythagorean distance formula gives the same distance between two points regardless of which Cartesian coordinate system calculations are made.

In Euclidean geometry, **a translation** is **a geometric transformation** that moves every point of a figure or

space by the same distance in a given direction.
A translation can also be interpreted as the addition of a constant vector to every point, or as **shifting the origin of the coordinate system**. In a Euclidean space, any translation is an isometry [4.2].

In mathematics, **a rigid transformation** (also called Euclidean transformation or Euclidean isometry) is a geometric transformation of a Euclidean space that **preserves the Euclidean distance between every pair of points** [4.3].

The two inertial reference systems that are considered within SR, when deriving LT, are two Cartesian coordinate systems that are linearly translated towards each other. This means that we can apply the Pythagorean distance formula to calculate the distance between two points.

This simple conclusion tells us that the distance between two points in S and S' is the same, **no length contraction takes place**!

4.3. Speed of light

In his work from 1864, Maxwell states that light is an electromagnetic phenomenon. The speed of light:

$c = 1/(\mu_0\varepsilon_0)^{1/2} = 299{,}792.458\ km/s$, approximately $300{,}000\ km/s$.
This speed, *c*, is the absolute speed of light, that is, the speed of light in light-bearing ether—in vacuum.
However, concerning the special theory of relativity, different inertial reference systems are discussed here.
They then refer all concepts to a reference system, even when one considers the speed of light.
Einstein formulated two postulates on which the SR is based. The following are obtained from [4.4].
1. The principle of relativity—the laws by which the states of physical systems undergoing change are not affected, whether these changes of state refer to one of the systems or both in a uniform translatory motion relative to each other.

2. The principle of invariant light speed—'... light is always propagated in space with a definite velocity [speed] *c*, which is independent of the state of motion of the emitting body'. In other words, light in a vacuum propagates with speed *c* (a fixed constant, independent of direction) in at least one system of inertial coordinates (**the "stationary system"**), regardless of the state of motion of the light source.

Thus, from the light postulate and Maxwell's theory, we can draw the following conclusions:
Light in vacuum propagates with speed *c*, a fixed

constant, independent of the direction, source motion, and observer motion!
Although this is true, they relate the speed of light to two reference systems when deriving LT, **which is incorrect**.

In [4.5], starting from two inertial reference systems, S and S', that move relatively each other at a constant speed, $v > 0$, S is considered stationary, while S' moves relative to S at a constant speed, $v > 0$.
A special case of 3 is used in [4.5], in which the light postulate is applied.
SC3: $x' = ct'$, $x = ct$; see notations in [4.6].
The light signal propagates in S according to the equation $x = ct$, which is correct because S is stationary with respect to the vacuum. This means that its absolute speed in space is zero.
However, S' moves at velocity $v > 0$ relative to S, which means that its absolute velocity in space, on the x-axis, is $v > 0$. This means that the equation for the light propagation in S', in SC3, $x' = ct'$, is incorrect! One can only write like this if S' is also stationary with respect to the vacuum.
This is the fundamental error that causes all the other inconveniences that result from the derivation of the LT. For more evidence one can consult [4.6], where the LT is proven to be applicable only when $v = 0$.
A similar error is committed in [4.7].

For inertial reference systems S, which are stationary relative to vacuum,

$$x = ct.$$

For inertial reference systems S', which are in motion relative to S, relative vacuum,

$$x' = ct'.$$

This is incorrect! Fig. 4.3.

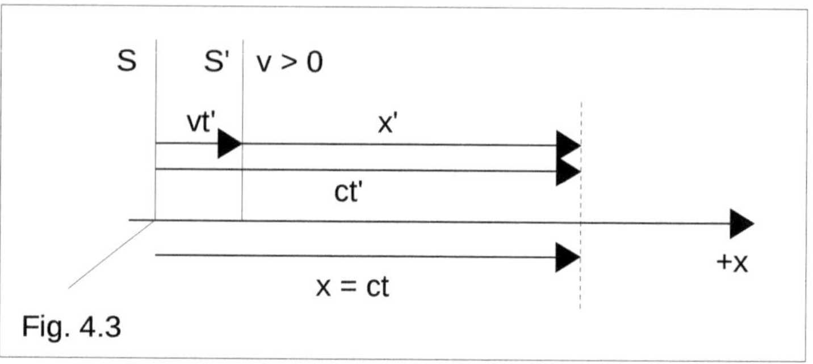

Fig. 4.3

When the two inertial reference systems are at the same point, $t = 0$ and $t' = 0$, a light signal arises in their origin. When time $t > 0$ and $t' > 0$, the light signal reaches a distance $\boldsymbol{ct = ct'}$. During this time, S' moves a distance of vt'. The distance between the S'-origin and the signal wavefront is denoted by x'.
From Fig. 4.3, we see that

$$ct' = vt' + x' \rightarrow x' = ct' - vt' = (c - v)t' \rightarrow$$
$$x' = (c - v)t'.$$

Therefore, in [4.7], the equation $x' = ct'$ should be replaced by the equation $x' = (c - v)t'$.
The minus sign indicates that S' and the wavefront of the signal move in the same direction. To delve into these issues further, one can consult [4.8].

4.4. Lorentz transformations and mathematics

In [4.5], a derivation of the LT is observed. A summary of this derivation is given below:
This starts from two inertial reference systems that move relatively each other at a constant speed, $v > 0$. These two reference systems are at the same point at the beginning of the experiment: $t = 0$ and $t' = 0$, Fig. 4.4.

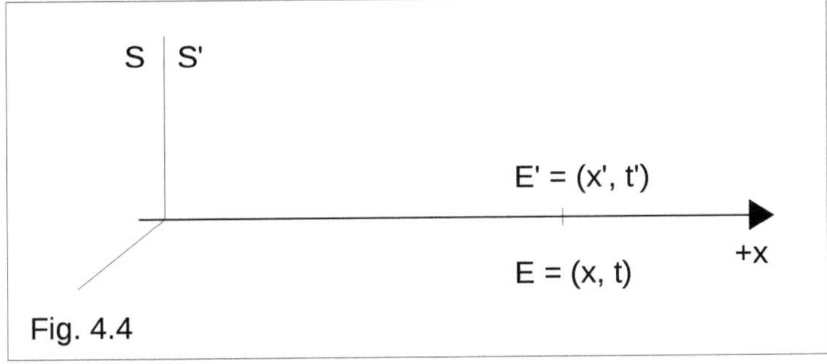

Fig. 4.4

We follow the reasoning and calculations from [4.5], pages 14–15.

Some notations are used to analyse this derivation, see [4.6].
An event that occurs in the two reference systems becomes

$E = (x, y, z, t)$ for the inertial reference system S,
$E' = (x', y', z', t')$ for the inertial reference system S'.

To simplify the calculations, we assume that the event occurs on the x-axis (x'-axis). The event then becomes

$E = (x, t)$ or $E' = (x', t')$.

In the beginning, we have two general linear equations, LE.
It is assumed that the transition between the coordinates in the two reference systems is linear.

LEx': $x' = Ax + Bt$,
LEt': $t' = Cx + Dt$,

where A, B, C, and D are constants.

To solve the above system of equations, three special cases are considered:

1) Event E' occurs in S'-origin. $E' = (x', t') = (0, t')$

2) Event E occurs in S-origin. $E = (x, t) = (0, t)$
3) The light postulate is applied, which states that the speed of light is the same in all inertial reference systems: $x = ct$ and $x' = ct'$ (we saw before that the equation $x' = ct'$ is used incorrectly, thus it must be replaced by $x' = (c - v)t'$).
From these, three pairs of special cases, SC, arise:

SC1: $x' = 0,\ x = vt,$
SC2: $x' = -vt',\ x = 0,$
SC3: $x' = ct',\ x = ct.$

These special cases are substituted in LEx' and LEt'.

LEx', SC1: $\rightarrow 0 = Avt + Bt,$
LEt', SC1: $\rightarrow t' = Cvt + Dt,$
LEx', SC2: $\rightarrow -vt' = Bt,$
LEt', SC2: $\rightarrow t' = Dt,$

LEx', SC3: $\rightarrow ct' = Act + Bt,$
LEt', SC3: $\rightarrow t' = Cct + Dt.$

Here, it is seen that these results show different relations between $t, t', A, B, C,$ and D.
Furthermore, the following is obtained:

From LEx', SC1: $\rightarrow 0 = Avt + Bt \rightarrow B = -Av$
From LEx', SC2 and LEt', SC2 $\rightarrow B = -Dv$

$\rightarrow$
$D = A$

From LEx', SC3 and LEt', SC3 and $B = -Av$, $D = A \rightarrow$
$C = -Av / c^2$

Finally, by replacing A with γ (gamma), the Lorentz transformations, LT, are obtained.

LTx': $x' = (x - vt)\gamma$,
LTt': $t' = (t - vx / c^2)\gamma$,
where $\gamma = 1 / (1 - v^2 / c^2)^{1/2}$ is called the Lorentz factor. End of summary.

An image is displayed when $t > 0$ and $t' > 0$.
From Fig. 4.5, we obtain the following:

$x = x' + vt$ or
$x' = x - vt$.

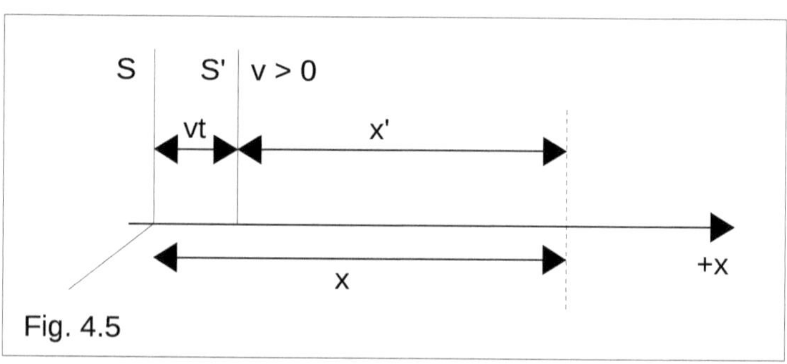

Fig. 4.5

This relationship is compared with LTx':

LTx' : $x' = (x - vt)\gamma$

Mathematics in Fig. 4.5: $x' = x - vt$

$\rightarrow \gamma = 1$

$\rightarrow v = 0.$

Every time we verify LT and apply conditions that reflect reality, a result that contradicts the original conditions is obtained. Here too, the LT applies only to $v = 0$.

A similar result is obtained if we consider the time t', which applies in S,' Fig. 4.6.

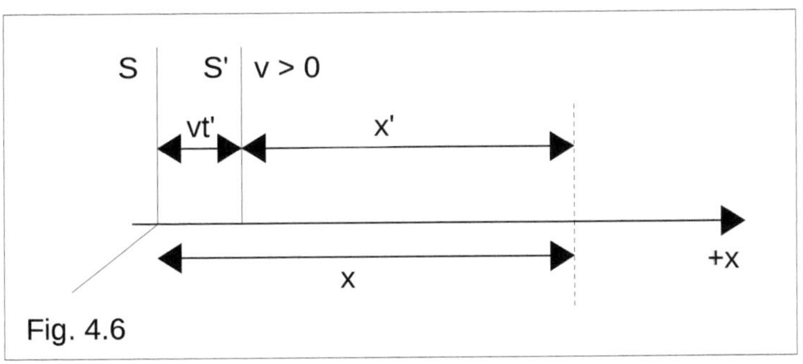

Fig. 4.6

Thus, $x = x' + vt'$.

See [4.9] where **the inverse Lorentz transformation** is obtained:

LTx: $x = (x' + vt')\gamma,$

LTt: $t = (t' + vx'/c^2)\gamma.$

The relationship from Fig. 4.6 is compared with LTx.

LTx: $x = (x' + vt')\gamma.$

Mathematical computation in Fig. 4.6: $x = x' + vt'$

$\rightarrow \gamma = 1$

$\rightarrow v = 0.$

4.5. Lorentz transformations when $v = 0$

Two reference systems are considered that are at a distance d from each other. They are stationary relative to each other, $v = 0$.

An event occurs on their common x-, x'-axis at a distance x from the S-origin.

This means that, according to mathematical principle, we have the following relationship between the coordinates of the event.

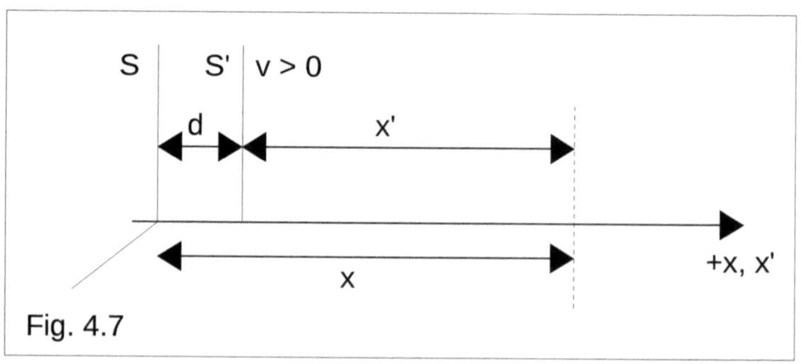

Fig. 4.7

Mathematical principle in Fig. 4.7: $x = x' + d$

The LT is verified by setting $v = 0$ in equation LTx', LTt', and γ.

LTx': $x' = (x - vt)\gamma$
LTt': $t' = (t - vx/c^2)\gamma$,

where $\gamma = 1/(1 - v^2/c^2)^{1/2}$ is called the Lorentz factor.

$\rightarrow \gamma = 1$
$\rightarrow$ LTx': $x' = x$
$\rightarrow$ LTt': $t' = t$.

From the mathematical principles in Fig. 4.7, we have

$$x = x' + d \rightarrow d = 0.$$

This means that the LT satisfies real situations only if the two reference systems are at the same point.
In this case, no transformations are required. In other words, the coordinates of the event in the two reference systems are identical.
According to the LT, two inertial reference systems that are at a distance $d > 0$ from each other and are stationary relative to each other cannot be obtained.
Then, the special theory of relativity is superfluous.
This means that SR is a theory that tries to alter the normal path of real events.

4.6. Lorentz transformations and basic definitions of their concepts

An inertial reference system S is considered, as shown

in Fig. 4.8. On the x-axis, an event $E = (x, t)$ occurs. The event is a short flash of light. S is stationary with respect to the vacuum.

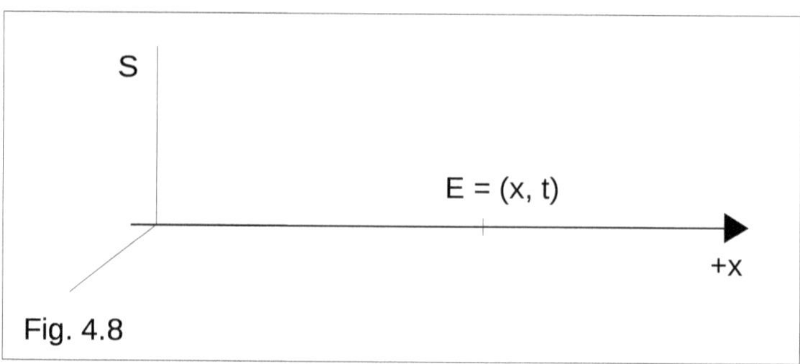

Fig. 4.8

To further use the coordinates of the event, we must obtain a clear definition of what these coordinates mean, how they are connected, and what they represent.
Starting with the x-coordinate, in this case, it is obvious that x represents the distance between the S-origin and the point on the x-axis where the event occurred, Fig. 4.9.
However, what does the t-coordinate represent?
It represents time, but what time?
Rather, the t-coordinate represents a moment, a pointer of a clock.
A watch is found in S-origin. Here, several questions arise.
How does the observer in S-origin know that an event

has occurred at a point x on the x-axis? What moment does the t-coordinate represent?
Is the observer in S-origin a human? How can a person determine the exact times?
One can know that an event has occurred only if one sees the light signal that occurs at point x.

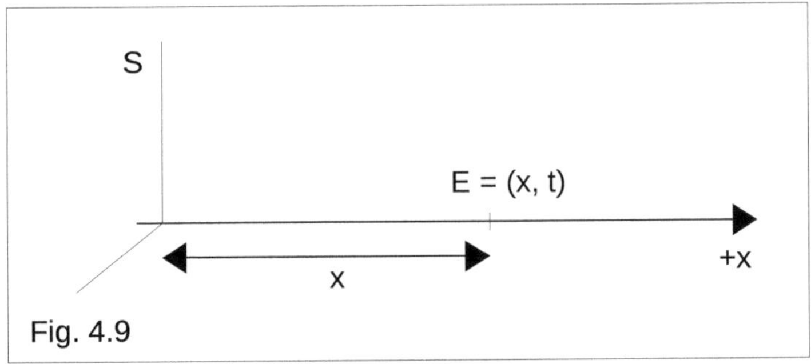

Fig. 4.9

The most realistic case could be that the watch in S-origin has a sensor that can read a light signal and register the moment when the light signal reaches the watch or its sensor.
The clock shows the time t_s when it receives the light signal.
Then, we can calculate the time showed by the clock when the event occurred.

$$t = t_s - x/c, \quad t_s = t + x/c.$$

Are these times the t-coordinates to represent?
If we think so, we get the following reasoning.

Based on this, we have the following definition for the coordinates of the event:

$$E = (x, t) = (x, x / c)$$

From this, it can be concluded that **the coordinates of an event cannot be determined if the distance to this event is not known**.
Here, a stationary reference system is considered.

SR is always about two reference systems, one is stationary and the other moves at a constant speed relative to the stationary reference system, with $v > 0$.
What does the coordinate x' represent? Is it the distance between the S'-origin and the point where the event occurs? Yes, but at what moment?
Is it the distance when the event occurs or the distance when S' becomes aware that the event occurred?
Physicists, especially those who defend the special theory of relativity, are challenged to determine the significance of the coordinates of the event in a reference system that is in absolute rest in space or in motion, and also in motion relative to a stationary reference system, as this is not done in the special theory of relativity.

4.7. Lorentz transformations and reality

My motto:
When studying physical phenomena, a mathematical model of these phenomena is created. In such a model, built-in physical laws are held together by mathematical tools.
If the description of the physical phenomenon is correct, the mathematical model is without contradictions and paradoxes.

We consider the two inertial reference systems S and S' again.
At the beginning of the experiment, both reference systems are at the same point. Then, their clocks are synchronised, $t = 0$ and $t' = 0$.
After a certain time, $t > 0$ and $t' > 0$, these two reference systems are at a distance > 0 from each other.

We focus on the distance between the two reference systems.
The distance between two points in two different reference systems was considered in Section 4.2.
The distance with respect to S is vt, whereas that with respect to S' is vt'. This means that,

for S: $d(S, S') = vt$
for S': $d(S, S') = vt'$

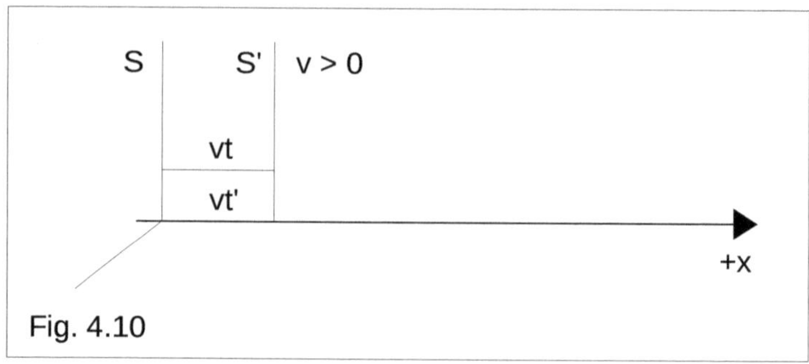

Fig. 4.10

But we have only one distance, one reality.

$\rightarrow$

$vt = vt' \rightarrow t = t'$.

Therefore, **there is no time dilation**.
If we apply this to the formula for time dilation, $t = t'\gamma$, we get

$\gamma = 1 \rightarrow$
$v = 0$.

This result also shows that the LT only applies to $v = 0$.

4.8. Lorentz transformations and light postulate

In section 4.3, we have addressed the problem of the light postulate. In [4.5], in SC3, the following conditions are used

SC3: $x' = ct'$ and $x = ct$.

I firmly claim here that the translation of the definition of the light postulate into mathematical relations is the following and nothing else:

$$ct' = ct$$

or that the speed of light has the same designation and the same value in both reference systems, c (we do not have c in S and c' in S').

The relation $ct' = ct$ is not the same as saying that $x' = ct'$ *and* $x = ct$!

The mathematical expression $ct' = ct$ says that we have the same speed of light in both reference systems. The mathematical expression $x' = ct'$ represents the distance between the S'-origo and the front of the light wave as a function of time t'.

The mathematical expression $x = ct$ represents the distance between the S-origo and the front of the light wave as a function of time t.

Because S is considered as a stationary reference system and S' as a moving reference system, these two are not equivalent and therefore the function of the distance between the origin and the front of the light wave **can not be the same**.

We have pointed out before that the relation $x' = ct'$ in

S' should be replaced by

$$x' = (c - v)\, t'.$$

What happens if we use this expression in the derivation of LT in [4.5]?
You start with two linear equations:

LEx': $x' = Ax + Bt$
LEt': $t' = Cx + Dt$

With a little simple mathematics, we get the corresponding inverse transformation

LEx: $x = (D/K)x' - (B/K)t'$
LEt: $t = -(C/K)x' + (A/K)t'$
where $K = AD - BC$.

In [4.5] the following special cases are used to determine the constants A, B, C and D.

SC1: $x' = 0,\ x = vt$
SC2: $x' = -vt',\ x = 0$
SC3: $x' = ct',\ x = ct$

We replace $x' = ct'$ with $x' = (c - v)t'$.

SC3: $x' = (c - v)t',\ x = ct$.

Now we make calculations:

LEx', SC1:

$x' = Ax + Bt$

$x' = 0$

$x = vt$

$\rightarrow$

$\boldsymbol{B = -vA}$

LEt', SC1:

$t' = Cx + Dt$

$x' = 0$

$x = vt \rightarrow$

$t' = (Cv + D)t$

LEx', SC2:

$x' = Ax + Bt$

$x' = -vt'$

$x = 0$

$\rightarrow$

$-vt' = Bt$

LEt', SC2:

$t' = Cx + Dt$

$x' = -vt'$

$x = 0 \rightarrow$

$t' = Dt$

From the last two results, $-vt' = Bt$ and $t' = Dt$ you get

$B = -vD \rightarrow$

$\boldsymbol{D = A}$.

LEx', SC3:

$x' = Ax + Bt$

$x' = (c - v)t'$

$x = ct \rightarrow$

$(c - v)t' = (Ac + B)t$

LEt', SC3:

$t' = Cx + Dt$

$x' = (c - v)t'$

$x = ct \rightarrow$

$t' = (Cc + D)t$

From the last two results, $(c - v)t' = (Ac + B)t$ and $t' = (Cc + D)t \rightarrow$

$(c - v)(Cc + D) = (Ac + B)$

We use previous results $B = -vA$ and $D = A \rightarrow$

$(c - v)(Cc + A) = Ac - Av \rightarrow$

$(c - v)(Cc + A) = (c - v)A \rightarrow$

$Cc + A = A \rightarrow$

$Cc = 0 \rightarrow$

$\boldsymbol{C = 0!}$

We have received $B = -vA$, $D = A$, $C = 0$ and

$K = AD - BC = A^2$ (In this case, when $x' = (c - v)t'$).

By replacing these in LEx', LEt' we get:

LTx': $x' = (x - vt)A$

LTt': $t' = At$

In [4.5] the value of A is determined by assuming that Lorentz transformations are symmetric and by replacing in the above LTx'and LTt'

x' with x,

t' with t,

x with x',

t with t'

v with $-v$

you get inverse LT

LTx: $x = (x' + vt')A$

LTt: $t = At'$

But we also get inverse LT from

LEx: $x = (D / K)x' - (B / K)t'$

LEt: $t = -(C / K)x' + (A / K)t'$

where $K = AD - BC = A^2$

(In this case, when $x' = (c - v)t'$).

By replacing the value of constants A, B, C, D and K with their values:

LEx: $x = (x' + vt')/A$
LEt: $t = t'/A$

Now we have two different expressions for inverse LT.
LTx: $x = (x' + vt')A$
LTt: $t = At'$
and
LEx: $x = (x' + vt')/A$
LEt: $t = t'/A$

We equate $t = At'$ and $t = t'/A \rightarrow$
$At' = t'/A \rightarrow$
$A = 1/A \rightarrow$
$A^2 = 1 \rightarrow$
$A = \pm 1$

All our experiments were done for cases there
$t > 0, t' > 0 \rightarrow$
$A > 0 \rightarrow$
$\mathbf{A = 1}$

For $A = 1$ we get
LTx': $x' = x - vt$
LTt': $t' = t$

LTx: $x = x' + vt'$
LTt: $t = t'$

These represent Galileo's transformations!
Clear, no weirdness!
We always get these transformations when we have made a mathematical model of the various special cases used in the derivation of LT and **when we use a correct model of reality**.

4.9. Conclusion

If we build our mathematical model correctly and apply mathematics, physics, and logic correctly, then the LT will never verify our model; therefore, the LT will never match real-life situations. Instead, by using the LT, we come to various contradictions and paradoxes. For further information on how the author has analysed various concepts within SR, the reader can consult [4.6, 4.8, 4.10–4.12].

4.10 References

[4.1] https://en.wikipedia.org/wiki/Euclidean_distance
[4.2] https://en.wikipedia.org/wiki/Translation_(geometry)
[4.3] https://en.wikipedia.org/wiki/Rigid_transformation
[4.4] https://en.wikipedia.org/wiki/Special_relativity
[4.5] Harris R 2008 *Special Relativity in Modern Physics*, chap. 2
[4.6] Slowak J 2020 *Special Relativity Is Nonsense*

[4.7] Einstein A 2006 *Den speciella och den allmänna relativitetsteorin*;
Första delen; Om den speciella relativitetsteorin; Swedish
[4.8] Slowak J 2022 *Light – The Absolute Reference in the Universe*
[4.9] https://en.wikipedia.org/wiki/Special_relativity#Lorentz_transformation_and_its_inverse
[4.10] Slowak J 2020 *Physics Essays: Mathematics Shows That the Lorentz Transformations Are Not Self-Consistent*
[4.11] Slowak J 2020 *SCIREA J. Phys.: Lorentz Transformations and Time Dilation Do Not Verify Reality*
[4.12] Slowak J 2020 *SCIREA J. Phys.: Lorentz Transformations - The Sound Versus The Light*

5. Alternative Lorentz transformations

We will build alternative Lorentz transformations. We will derive **new transformations** between the two reference systems. It will be easy to compare these two variants of Lorentz transformations. We will show that if all the steps taken during the derivation apply the existing mathematics, logic and physics, our transformations will be flawless, contradiction free! We follow the same steps, the same way of thinking as one do in [5.1].

5.1 Our thought experiments

Imagine a highway, perfectly straight and perfectly horizontal. On this highway, we mark a point where an observer S is located. An additional observer, S', is at the same point at the beginning of each thought experiment (in our case we can do these experiments for real). The observer S' moves at constant speed $v > 0$ to the right in our model. We decide that $v = 2\ m / s$.

The two observers exchange information using a Tesla car that moves during our experiments at a constant speed $w = 20\ m / s$.

An event that occurs in our reality will be considered

as a point in the two 2-dimensional reference systems:

(x, t) for S

(x', t') for S'

where x, x' is the coordinate of space and t, t' is the coordinate of time.

We will try to determine **two linear transformations** (equations) between (x, t) and (x', t') and vice versa.

We denote them by LEx' and LEt':

LEx': $x' = Ax + Bt$

LEt': $t' = Cx + Dt$

With a little simple mathematics, we get the corresponding **inverse transformation**

LEx: $x = (D / K)x' - (B / K)t'$

LEt: $t = - (C / K)x' + (A / K)t'$

where $K = AD - BC, K \neq 0$.

These two systems of equations are equivalent.

To determine the constants A, B, C and D, we perform two thought experiments and name them special cases, SC.

We consider two inertial reference systems, S and S', two 2-dimensional coordinate systems. Their x-axis and x'-axis coincide on the same line.
When S and S' are at the same point, $t = 0$, $t' = 0$.

5.2. SC1

At the beginning of this experiment, S and S' are at the same point. A Tesla car is moving at a constant speed, $w > 0$, **from the right** towards S and S'.

Fig. 5.1

After a time, $t' > 0$, Tesla, on its way to S, meets S'.

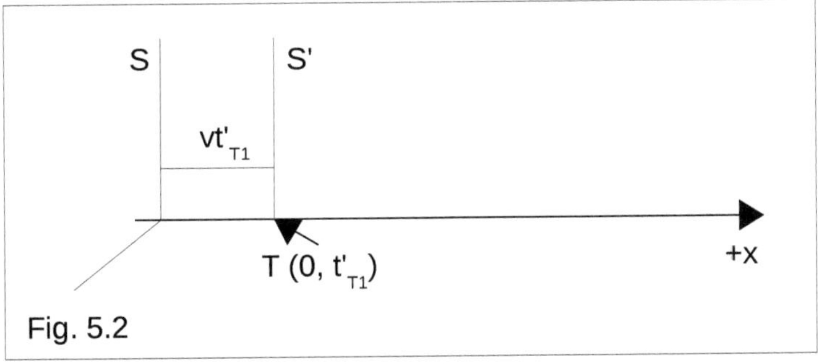

Fig. 5.2

At this moment S' reads time t'_{T1} and considers that the event has occurred in its origin, $x'_{T1} = 0$. The event when T meets S' becomes:

$$(x', t') = (0, t'_{T1}).$$

It is obvious that the distance between S and S', at this moment, is vt'_{T1}!

After this, the car continues on to S and when it reaches this observer, S reads the time t_{T1}. What value does t_{T1} have?
t_{T1} is t'_{T1} plus the time the car needs to drive the distance vt'_{T1}.

$$t_{T1} = t'_{T1} + vt'_{T1}/w \rightarrow t_{T1} = t'_{T1}(1 + v/w) \rightarrow$$

$$\mathbf{t_{T1} = t'_{T1}(w + v)/w}$$

Then S can calculate the time when the event occurred in S'-origo.

$$\mathbf{t'_{T1} = t_{T1}w/(w + v)}$$

and can then also calculate the distance to the point where the event occurred:

$$x = vt'_{T1} \rightarrow$$

$$x = t_{T1}wv/(w + v)$$

Now we have the coordinates of the event for both S' and S

$\boldsymbol{(x', t') = (0, t'_{T1})}$

$\boldsymbol{(x, t) = (t_{T1}wv/(w + v), t_{T1}w/(w + v))}$

We replace these coordinates in LEx' and LEt' to determine A, B, C and D.

From LEx', (x', t') and (x, t) we get

LEx': $x' = Ax + Bt$

$(x', t') = (0, t'_{T1})$

$(x, t) = (t_{T1}wv / (w + v), t_{T1}w / (w + v))$

$\rightarrow$

$0 = At_{T1}wv / (w + v) + Bt_{T1}w / (w + v) \rightarrow$

$0 = Av + B \rightarrow$

$\boldsymbol{B = - Av}$

From LEt', (x', t') and (x, t) we get

LEt': $t' = Cx + Dt$

$(x', t') = (0, t'_{T1})$

$(x, t) = (t_{T1}wv / (w + v), t_{T1}w / (w + v))$

$\rightarrow$

$t_{T1}w / (w + v) = Ct_{T1}wv / (w + v) + Dt_{T1}w / (w + v) \rightarrow$

$1 = Cv + D \rightarrow$

$\boldsymbol{C = (1- D)/v}$

We get the same value for B and C if we use

$(x', t') = (0, t'_{T1})$

$(x, t) = (vt'_{T1}, t'_{T1})$.

5.3. SC2

At the beginning of this experiment, S and S' are at the same point. A Tesla car is moving at a constant speed, $w > 0$, **from the left** towards S and S', Fig. 5.3.

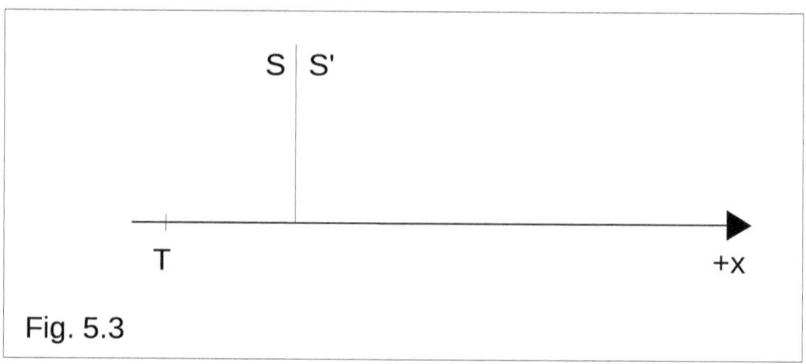

Fig. 5.3

After a time, $t > 0$, Tesla, on its way to S', meets S, Fig 5.4.

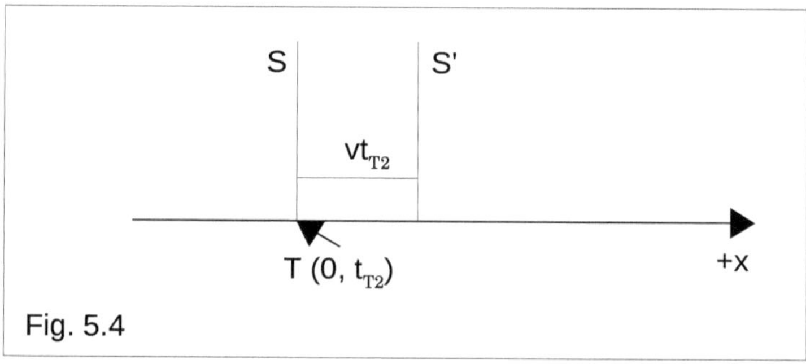

Fig. 5.4

When the car passes S, the observer in S reads the time t_{T2}. It is considered that the an event occurred in S-origo.

$(x, t) = (0, t_{T2})$.

It is obvious that the distance between S and S', at this moment, is vt_{T2}!

After this, the car continues on to S'. But as the car approaches S', this reference system manages to go an extra distance.
The observer in S' reads the time t'_{T2}, Fig. 5.5.
The distance between S and S' at this moment is x'.
We see that distance $d(SS') = d(x')$

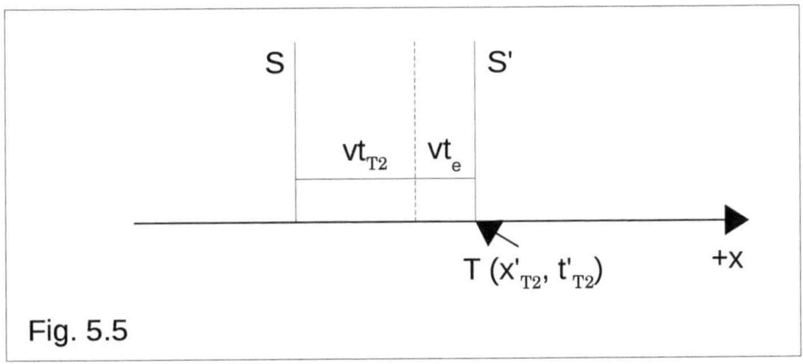

Fig. 5.5

$$d(x') = vt'_{T2} \rightarrow$$
$$d(x') = vt_{T2} + vt_e \rightarrow$$
$$t'_{T2} = t_{T2} + t_e$$

but we also see that

$$d(x') = wt_e$$

This is the distance that the car moves between S and

S'. From here we get

$vt_{T2} + vt_e = wt_e \rightarrow$
$vt_{T2} = t_e(w - v) \rightarrow$
$t_e = t_{T2}v/(w - v) \rightarrow$

From
$t'_{T2} = t_{T2} + t_e$ and
$t_e = t_{T2}v/(w - v) \rightarrow$
$t'_{T2} = t_{T2} + t_{T2}v/(w - v) \rightarrow$
$\boldsymbol{t'_{T2} = t_{T2}w/(w - v) \rightarrow}$
$\boldsymbol{t_{T2} = t'_{T2}(w - v)/w}$

Now we have the coordinates of the event for both S' and S for SC2
$\boldsymbol{(x, t) = (0, t_{T2})}$
$\boldsymbol{(x', t') = (-vt'_{T2}, t'_{T2}(w - v)/w)}$

We have the minus sign because x' is measured to the left, towards the negative part of the x-axis, x'-axis. We replace these coordinates in LEx' and LEt' to determine *A, B, C* and *D*.

From LEx', (x', t') and (x, t) we get
LEx': $x' = Ax + Bt$
$(x, t) = (0, t_{T2})$
$(x', t') = (-vt'_{T2}, t'_{T2}(w - v)/w)$

$\rightarrow$

$-vt'_{T2} = A*0 + Bt'_{T2}(w-v)/w \rightarrow$

$\boldsymbol{B = -wv/(w-v)}$

From LEt', (x', t') and (x, t) we get

LEt': $t' = Cx + Dt$

$(x, t) = (0, t_{T2})$

$(x', t') = (-vt'_{T2},\ t'_{T2}(w-v)/w)$

$\rightarrow$

$t'_{T2}(w-v)/w = C*0 + D\,t'_{T2}(w-v)/w \rightarrow$

$\boldsymbol{D = 1}$

5.4. Merger of results

From these two thought experiments we obtained the following relations for the constants *A, B, C* and *D*.

$B = -Av$

$C = (1-D)/v$

$B = -wv/(w-v)$

$D = 1$

$\rightarrow$

$A = -B/v \rightarrow$

$\boldsymbol{A = w/(w-v)}$

$C = (1-D)/v \rightarrow$

$\boldsymbol{C = 0}$

We have seen in section 5.1 that the inverse

transformation has the form

LEx: $x = (D/K)x' - (B/K)t'$
LEt: $t = -(C/K)x' + (A/K)t'$
where $K = AD - BC$.

When we calculate the value of the expression $AD - BC$ we get
$K = A*1 - B*0 = A$
$\boldsymbol{K = A}$

Now we can write the two new transformations between coordinate systems for S and S'. From
LEx': $x' = Ax + Bt$
LEt': $t' = Cx + Dt$

and A, $B = -Av$, $C = 0$, $D = 1$
→
NTx': $x' = A(x - vt)$
NTt': $t' = t$

We replace A, B, C, D and K in
LEx: $x = (D/K)x' - (B/K)t'$
LEt: $t = -(C/K)x' + (A/K)t'$
→
NTx: $x = (1/A)x' + vt'$
NTt: $t = t'$

We have obtained two pairs of new transformations between the coordinates of the two inertial reference systems:

NTx': $x' = A(x - vt)$
NTt': $t' = t$

NTx: $x = (1/A)x' + vt'$
NTt: $t = t'$

Our two events from our two special cases are:

SC1 $(x', t') = (0, t'_{T1})$
$(x, t) = (t_{T1} wv / (w + v), t_{T1} w / (w + v))$

SC2 $(x, t) = (0, t_{T2})$
$(x', t') = (-vt'_{T2}, t'_{T2}(w - v) / w)$

But we also have the relationship between t and t' in each experiment:

SC1 $t_{T1} = t'_{T1}(w + v) / w$
$t'_{T1} = t_{T1} w / (w + v)$

SC2 $t'_{T1} = t_{T1} w / (w - v)$
$t_{T1} = t'_{T1}(w - v) / w$

In [5.1] the value of A is determined by assuming that Lorentz transformations are symmetric and by replacing in LTx' and LTt'

x' with x,
t' with t,
x with x',
t with t'
v with $-v$

We do the same in our two equations

NTx': $x' = A(x - vt)$
NTt': $t' = t$

and we get so-called inverse equations

NTx: $x = A(x' + vt')$
NTt: $t = t'$

5.5. Verification of calculations

We verify the following two systems of equations

NTx': $x' = A(x - vt)$
NTt': $t' = t$
$\rightarrow$
NTx: $x = A(x' + vt')$
NTt: $t = t'$

against the two thought experiments we did:

Because we have $t = t' \rightarrow$

NTx': $x' = A(x - vt)$
NTx: $x = A(x' + vt)$

Then we only verify for the event

SC1: $(x', t') = (0, t'_T)$

SC2: $(x, t) = (0, t_T)$

and as we can write them now in the following way

SC1: $(x', t') = (0, t)$

SC2: $(x, t) = (0, t)$

From

NTx': $x' = A(x - vt)$

NTx: $x = A(x' + vt)$

SC1: $(x', t') = (0, t)$

→

$0 = Ax - Avt$

$x = A(0 + vt)$

→

$0 = x - vt$

$x = Avt$

→

$x = vt$

$x = Avt$

→

$Avt = vt$

→

$A = 1$

From

NTx': $x' = A(x - vt)$

NTx: $x = A(x' + vt)$

SC2: $(x, t) = (0, t)$

→

$x' = A(0 - vt)$

$0 = A(x' + vt)$

→

$x' = -Avt$

$0 = x' + vt$

→

$x' = -Avt$

$x' = -vt$

→

$-Avt = -vt$

→

$A = 1$

So, two systems of equations became as follows:

NTx': $x' = x - vt$

NTt': $t' = t$

→

NTx: $x = x' + vt'$

NTt: $t = t'$

These are Galilean transformations!

There we are again: if we apply the mathematics, physics and logic correctly, no thought experiment can lead to Lorentz transformations. They don't verify reality!

But what if $A = 1$? Before we had

$$A = w / (w - v) \rightarrow$$
$$w / (w - v) = 1 \rightarrow$$
$$w = w - v \rightarrow$$
$$\boldsymbol{v = 0}$$

Again we get the result that LT only applies to $v = 0$. Why do we always get this result?

The reason is that we are trying to build linear transformations between S and S'.

Such transformations **do not exist** between S and S' if we use as the carrier of the message between these two reference systems light signals (or a Tesla car),
a carrier of the message which has a speed greater than the relative speed between S and S'.
In our case, a carrier of the message is a car with the speed

$$w = 20\ m / s$$

while the relative velocity between S and S' is

$v = 2\ m/s$

In any case, the speed of the carrier of the message must be greater than the relative speed between the two reference systems otherwise they cannot communicate with each other!

The transition from one reference system to another depends on how these two inertial reference systems move relative to each other and especially from which direction the light signal moves towards the reference system in motion, see [5.3].

Transformations between two coordinate systems in relative motion are NOT symmetrical!

5.6. Conclusions

If you apply physics, mathematics and logic correctly, you cannot create linear transformations between two inertial reference systems.
Not, if you use the two thought experiments we did or those used in [5.1].

All my verifications of Lorentz tarnsformations in SR give the conclusion that Lorentz tarnsformations only applies to $v = 0$! Therefore, my conclusion in all my research ends with that

Special Relativity is nonsense!

When the carrier of the information between the two observers comes from the right (as it approaches S' from the front), the conversion factor is $(w + v)/w$. When the carrier of the information between the two observers comes from the left (as it approaches S' from behind), the conversion factor is $(w - v)/w$. This does not mean that we have some time dilation. This means that the value for time coordinate in one reference system can be calculated using the value for time coordinate in the other reference system.

The time in the two reference systems runs at the same rate!

Think about how we did our two thought experiments. Both **distance** and **time** we use are **mathematical quantities**. We used the math to calculate them!

We have used current mathematics, current logic, and current classical physics!

Note that there are so many transformations between S and S' how many definitions of *(x, t)* and *(x', t')* there are!

The mathematical model you create MUST verify reality! No paradoxes, please!

Quote:
"At the same time, these are the worlds we have the hardest to understand, worlds where illustrative models deceive us and we find paradoxes. But there is only one world and it has no paradoxes. Only our models that hold paradoxes "
[Vid skiljevägen : essäer om människan och hennes framtid; Ulf Sinnerstad; 2006, Swedish]

Think this through carefully!

References

[5.1] *Special Relativity in Modern Physics;* Chapter 2; Randy Harris; 2008
[5.2] *Special Relativity is Nonsense*; third edition; Jan Slowak; 2020
[5.3] *Light - the absolute reference in the universe*; fourth edition; Jan Slowak; 2022

6. Lorentz's transformations are the culprit in the drama

Lorentz's transformations are the basis of the whole special theory of relativity, SR. What are they?

Lorentz's transformations:

LTx': $x' = (x - vt)\gamma$,

LTt': $t' = (t - vx/c^2)\gamma$,

where $\gamma = 1/(1 - v^2/c^2)^{1/2}$ is called the Lorentz factor.

Inverse Lorentz's transformations:

LTx: $x = (x' + vt')\gamma$,

LTt: $t = (t' + vx'/c^2)\gamma$,

Both apply at the same time, they are equivalent.

How consistent are these equations? I have made calculations to derive these equations dozens of times, either followed derivations from different university courses or just tried to find where the error lies. Something was not right. SR seemed wrong but how would one prove mathematically that it is wrong. Why mathematical proof? Because SR is primarily a theoretical construction, a mathematical construction!

Lorentz transformations own a method of transitioning from one reference system to another. A method of calculating the coordinates of the event in one reference system using the coordinates of the other reference system.

The two reference systems are considered symmetrical.
We assume that S is in absolute rest in space and S' moves relative to S towards the positive part of the x-axis.
This is symmetrical with: S' is at absolute rest in space and S moves relative to S' towards the negative part of the x-axis.

This principle of symmetry is widely used in SR. It's actually the case that without it you will not get Lorentz transformations!
This is also done in [6.1].
It is from Lorentz's transformations that one derives the time dilation.

S is at absolute rest, S' moves at speed $v > 0$ relative to S. You come to the relation

$$t' = t\gamma$$

S' is at absolute rest, S moves at speed $v > 0$ relative to S. You come to the relation

$t = t'\gamma$

From here you get

$t' = t'\gamma\gamma \rightarrow$

$\gamma^2 = 1 \rightarrow$

$\gamma = \pm 1$

Gamma, γ, is always positive $\rightarrow$

$\gamma = 1 \rightarrow$

$v = 0$

But we also get that $t = t'$!

References

[6.1] *Special Relativity in Modern Physics;* Chapter 2; Randy Harris; 2008

7. Analysis of k-calculus from Introducing Einstein's Relativity by Ray d'Inverno

We will analyze some aspects of the concepts from SR. We will show that no matter what method and model you use, you will always come across a contradiction. The contradiction is obtained if one carefully verifies the model with the reality, the physics, the mathematics, and the logic. We follow the book [7.1], chapter 2.

7.2.1 Model building

"The activity consists of constructing a mathematical model which we hope in some way capture the essentials of the phenomena we are investigating."

Yes, this is the most important moment in the explanation of a physical phenomenon.
If the mathematical model is done correctly, if it correctly reflects the physical phenomenon, then the following calculations, conclusions and results should not contradict either the model, the existing mathematical or physical laws.

7.2.2 Historical background

Here the author of [7.1] goes through some of the steps

made by different physicists, researchers, which ultimately resulted in the creation of the special theory of relativity. We mention some of them:

- 1865, James Clark Maxwell: theory of electromagnetism; light-bearing ether
- 1887, Michelson - Morley experiment; negative result
- 1904, Hendrik A. Lorentz: Lorentz transformations; Lorentz factor
- 1905, Albert Einstein: The theory of special relativity

Author Ray d'Inverno says:
"In fact, the essence of the special theory of relativity is conteined in the Lorentz transformations."
This is true but it is from these transformations that the contradictions emerge!

7.2.3 Newtonian framework

Here they talk about events, about space-time diagrams, world-line, observers.
We show our own picture of this.

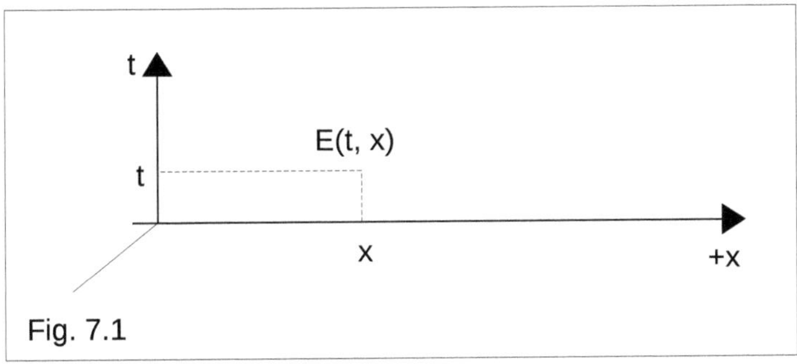

Fig. 7.1

We depict an event E that takes place at the time *t* on the t-axis and at the point *x* on the x-axis. Say that this figure represents an inertial reference system S. The point, at which the t-axis and the x-axis intersect, represents the origin of the reference system.

7.2.4 Galilean transformations

"N1: Every body continues in its state of rest or of uniform motion in a straight line unless it is compelled to change that state by forces acting on it."
This is logical, I see no problem with this statement.

The author of the book [7.1] says:
"Thus, there exists a privileged set of bodies, namely those not acted on by forces."
I do not think this conclusion can be drawn from the N1.

7.2.5 The principle of special relativity

Here they address, among other things:
"Many fundamental principles of physics are statements of impossibility, and the above statement of the relativity principle is equivalent to the statement of the impossibility of deciding, by performing dynamical experiments, whether a body is **absolutely in rest** or in uniform motion."

I argue that one can determine if an object is at

absolute rest or if it is moving at a constant speed. This is shown in the book [7.2].

Furthermore, reference is made to the first postulate Principle of special relativity: All inertial observers are eqvivalent.
I'm just asking the following question: if they are equivalent then why are their clocks ticking differently? This is a legitimate question!

7.2.6 The constancy of the velocity of light

Second postulate of SR, constancy of velocity of light: The velocity of light is the same in all inertial systems.

I accept this, how else? Maxwell has come to this conclusion in his work.
If $c = 1/(\mu_0\varepsilon_0)^{1/2}$ then c is not dependent on anything other than the properties of the medium:

the permittivity of free space, ε_0
the permeability of free space, μ_0.

The speed of light is one thing and **the relative speed** between two objects or between one object and the wavefront of the light signal is another.

7.2.7 The *k*-factor

In the figures and models here, one takes $c = 1$
(c = 1 light-second / 1 second).

In normal cases, $c = 299{,}792.458\ km/s$.
What does it mean?
This means that, in the 2-dimensional space-time diagram, you have the same "unit of length" both on the x-axis and on the t-axis. e.g. if you use on the t-axis as unit *1 (one) second* then the unit on the x-axis becomes *1 (one) light-second*.
Light-second, light-year are units of length.

But if you set $c = 1$, the corresponding conversion must also be made for v. Say we approximate c to $300{,}000\ km/s$ and the Earth's speed around the Sun to $30\ km/s$. Then these two speeds in the model from [7.1] will be as follows:

$c = 1$ *light-second / second*
$v = (1/10{,}000)$ *light-second / second*

A legitimate question: why complicate it, what were the purposes of creating this k-calculus model?

In such models, the world-line of a point from the wavefront of light will be at an angle to the x-axis and the t-axis of $\pi/4 = 45°$. See the next figure.
The line WL represents world-line for a light signal starting from the origin of the reference system. t_0 is time and is measured in e.g. seconds, $t_0 = 1$ *second*.
x_0 is the distance on the x-axis between the origin of the reference system and the point where the light

front reaches in the meantime t_0.
This distance is $x_0 = 1$ *light-second.*

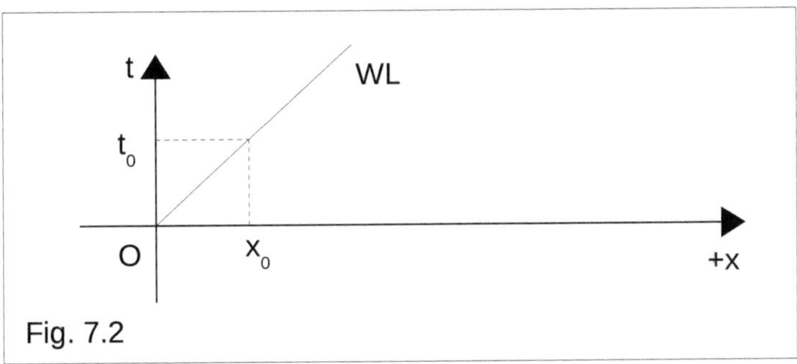

Fig. 7.2

Remember that all events occur on the x-axis, that even the light signal we observe moves on the x-axis!

An inertial reference system considered within SR moves at a speed less than that of light, $v < 1$, this means that the world-line of such a system will be at an angle less than 45° to the t-axis, Fig. 7.3.

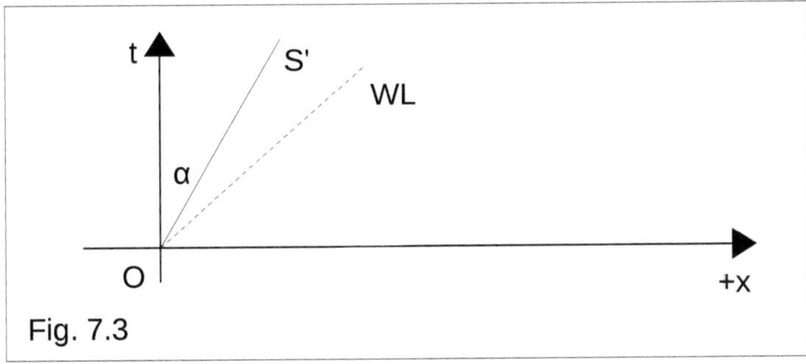

Fig. 7.3

We also show what it would look like if S' is at rest relative to S, $v = 0$.

Then the angle of world-line of S' to the t-axis will also be zero, they become parallel, Fig. 7.4.

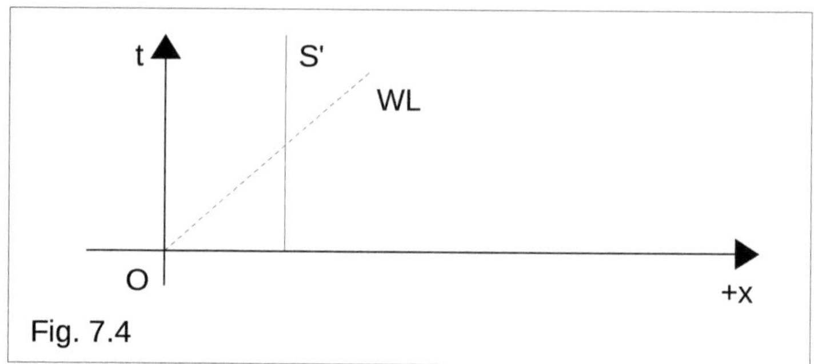

Fig. 7.4

The world-line of the reference system with the origin at the point O becomes a line that coincides with the t-axis.

Quote from [7.1]:
"Let us assume we have two observers, *A* at rest and *B* moving away from *A* with uniform (constatnt) speed."

This approach is used in almost all situations when treating SR, doing derivation of LT.
It is said that *A* is at rest. But I ask a question here that is very important:

Relatively what is *A* at rest? Can proponents of SR provide an example of such a reference system?

You can have this set of reference systems to investigate a physical phenomenon but not when using light signals!

Quote from [7.1]:
"In fact, there is a hidden assumption here, since how do we know that *B*'s world-line will be a straight line as indicated in the diagram?"

There is no assumption, if *B* moves with constant velocity then results from the model, from the geometry, that *B*'s world-line **is** a straight line.

What did the author of [7.1] want to hide, question here?

7.2.8 Relative speed of two inertial observers

We consider two inertial reference systems, S and S'. S' moves (to the right in the figure) at a constant speed $v > 0$ relatively S.

In this section, the following **thought experiments** are performed. At time t_0 a light signal is sent from S to S'. When the light signal reaches S', the signal is sent back to S. The world-line of the reflected light signal becomes a line that is symetrical with WL relative to the t-axis. The angle between these two world-lines will then be 90°.

We mark a number of points on the t-axis and on the x-axis.
Every pair of these points form distances and we will calculate their length. They also form some right-angled triangles and we will calculate the length of their sides.
Here are the calculations for some of the distances in Fig. 7.5:

Distance between two points P and Q, we denote by d(PQ).
- the distance between T_0 and T_0' is the distance that S' moved during the time t_0.

$$d(T_0T_0') = x_0 = vt_0$$

- the distance between P and P' is the distance between S and the point where S' is when the light signal arrives to it.

$$d(PP') = x_1 = vt_1$$

We calculate the time t_1 based on the Fig. 7.5.
The time interval $t_1 - t_0$ is the time the light signal from S needs to reach S'.

$x_1 = c(t_1 - t_0)$; We have $c = 1 \rightarrow \boldsymbol{x_1 = t_1 - t_0} \rightarrow$
$vt_1 = t_1 - t_0 \rightarrow t_0 = t_1 - vt_1 = t_1(1 - v) \rightarrow t_0 = t_1(1 - v)$
$\rightarrow$
$\boldsymbol{t_1 = t_0/(1 - v)}$

This reasoning, to compare distances that during the same time are passed both by the light signal and by the reference system in motion, I have not seen in any literature, only in [7.2-7.4].
This is the key to finding contradictions in the models that deal with the derivation of LT.

In the triangle OPP' we have the following relations. We denote the angle between OP and OP' by α. Then we have:

$tan\,\alpha = d(PP')\,/\,d(OP) = x_1\,/\,t_1 = v$ (tan = tangent)

This angle represents the slope of world-line for S' relative to the t-axis.

$d(OP')^2 = d(PP')^2 + d(OP)^2$ (Pythagorean theorem)
$\rightarrow$
$d(OP')^2 = (vt_1)^2 + t_1^2 \rightarrow \boldsymbol{d(OP')^2 = t_1^2(1 + v^2)}$

In the triangle $PP'T_0$ is $d(PP') = d(PT_0)$ because the opposite angles are 45°. The same goes for the triangle $PP'T_2$, $d(PP') = d(PT_2)$.

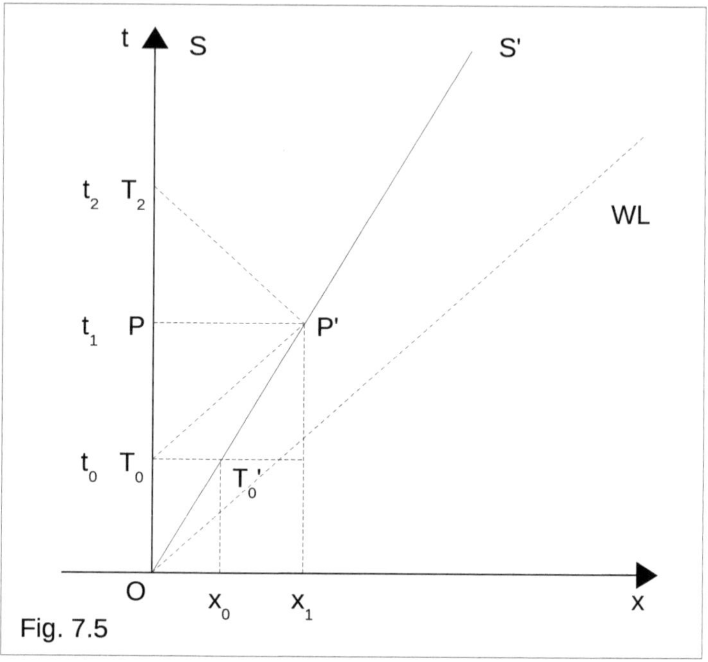

Fig. 7.5

All these distances are equal to $\boldsymbol{t_1 - t_0}$.
See calculations above.

Now we can calculate t_2, when the light signal returns to S.

$$t_2 = d(T_2P) + d(PT_0) + d(T_0O) = (t_1 - t_0) + (t_1 - t_0) + t_0 \rightarrow$$

$$t_2 = (t_1 - t_0) + (t_1 - t_0) + t_0 \rightarrow$$

$$t_2 = 2t_1 - t_0 = 2t_0 / (1 - v) - t_0 = (2t_0 - t_0(1 - v)) / (1 - v) = (2t_0 - t_0 + vt_0)) / (1 - v) \rightarrow$$

$$\boldsymbol{t_2 = t_0(1 + v)/(1 - v)}$$

We see that we can calculate all the distances between the different points formed in the model.
These distances depend **only** on

c – the speed of light, $c = 1$

v – the speed of S' relative to S, $0 < v < 1$

t_0 – the time when the light signal is transmitted from S to S'

We don't need to make any other assumptions!
We summarize:

$$\boldsymbol{t_1 = t_0/(1 - v)}$$

$$\boldsymbol{t_2 = t_0(1 + v)/(1 - v)}$$

The factors $(1 - v)$ and $(1 + v)$ occur abundantly in the book [7.2], although where the author uses c, the speed of light. There, they become equal to $(c - v)$ and $(c + v)$.

In the book [7.1] that we analyze, one denotes

$$k = ((1 + v) / (1 - v))^{1/2} \rightarrow k^2 = (1 + v) / (1 - v) \rightarrow$$

$$t_2 = k^2 t_0$$

It is the same relation as stated in [7.1]. Also the calculation that
$(k^2 - 1) / (k^2 + 1) = v$ is correct.

But the **assumption** that the time in S' is proportional to the time in S is not correct!

In the book [7.1] it is stated that the distance between the points O and P' is kt_0.

$$d(OP') = kt_0$$

We have seen before that $d(OP')^2 = t_1^2(1 + v^2)$.
We replace t_1 with $t_0/(1 - v) \rightarrow$

$$t_0^2(1 + v^2)/(1 - v)^2 = k^2t_0^2 \rightarrow (1 + v^2)/(1 - v)^2 = (1 + v)/(1 - v) \rightarrow$$
$$(1 + v^2) = (1 + v)(1 - v) \rightarrow 1 + v^2 = 1 - v^2 \rightarrow 2v^2 = 0 \rightarrow$$
$$\boldsymbol{v = 0}$$

This represents a contradiction to the original condition that S' moves at a speed *v > 0* relative to S.
We got this contradiction from **the only assumption** in our mathematical model, the assumption that $d(OP') = kt_0$, that the time in S' is proportional to the time in S with the factor k!

Why do you do that? You create a mathematical model, you use it to a certain point, but you do not pursue thinking. Why make a **assumption** here?
All distances can be calculated from the model!

I can not believe that the researchers who created k-calculus did not see that the OP' can be calculated from the triangle OPP' and that its length is

$d(OP') = t_1(1 + v^2)^{1/2}$.

7.2.11 The clock paradox

We make our own figure here too, to be able to explain better.

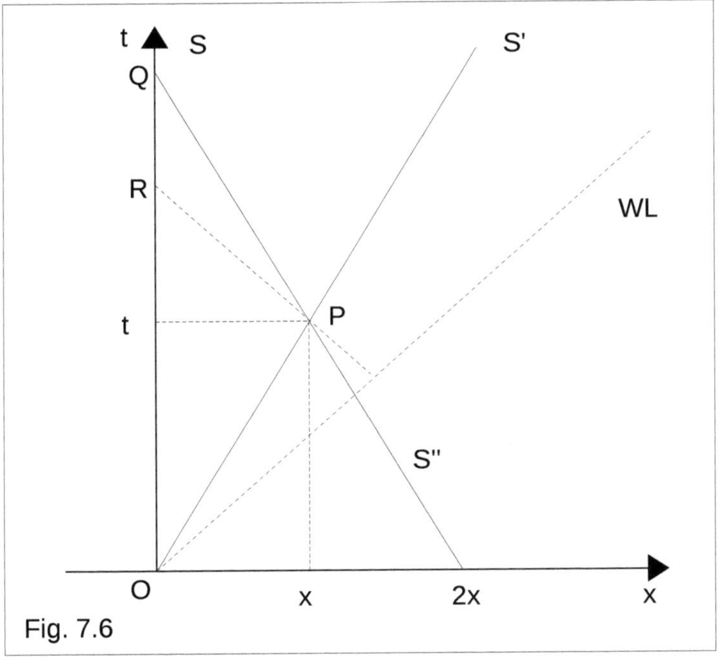

Fig. 7.6

The reference system S' moves to the right in the figure at speed v relatively S. When S' is in O, a third reference system S'' starts at the point $2x$ from O. S'' moves to the left at speed v relatively S. After time t, S' and S'' hit togather at the point P (at the point x on the x-axis). Then they send a light signal to S.

This signal has a world-line that is perpendicular to the WL.

Now we have all the parameters in place. Based on the figure, we can now calculate all distances.
From Fig. 7.6 we have

$x = vt; d(tP) = vt; d(Rt) = d(tP) \rightarrow$

$R = vt + t \rightarrow \boldsymbol{R = t(1 + v)}$

$d(OP) = T = (t^2 + (vt)^2)^{1/2} = t(1 + v^2)^{1/2} \rightarrow$

$\boldsymbol{T = t(1 + v^2)^{1/2}}$

If we make the assumption that $R = kT$ as one do in [7.1], we get a contradiction!

$R = kT \rightarrow t(1 + v) = kt(1 + v^2)^{1/2} \rightarrow (1 + v) =$
$k(1 + v^2)^{1/2} \rightarrow$

$(1 + v)^2 = k^2 (1 + v^2) \rightarrow (1 + v)^2 =$
$(1 + v)(1 + v^2)/(1 - v) \rightarrow$

$(1 + v)(1 - v) = (1 + v^2) \rightarrow 1 - v^2 = 1 + v^2 \rightarrow \boldsymbol{v = 0}$

This represents a contradiction to the original condition that S' moves at a speed $v > 0$ relative to S (the same calculations as before)**.**

We see clearly from the figure that (all lengths in the figure are only dependent on the time t when S and S' meet)

$\boldsymbol{Q = 2t}$

The author of [7.1] says that $Q = (k + k^{-1})T$, where $k = ((1 + v)/(1 - v))^{1/2}$.

Calculation:

$2t = (k + k^{-1})t(1 + v^2)^{1/2} \rightarrow$

$2/(1 + v^2)^{1/2} = (1 + v)^{1/2}/(1 - v)^{1/2} + (1 - v)^{1/2}/(1 + v)^{1/2} \rightarrow$

$2/(1 + v^2)^{1/2} = ((1 + v) + (1 - v))/(1 + v)^{1/2}(1 - v)^{1/2} \rightarrow$

$2/(1 + v^2)^{1/2} = 2/(1 - v^2)^{1/2} \rightarrow 1 + v^2 = 1 - v^2 \rightarrow$

$\boldsymbol{v = 0}$

Quote from [7.1]:
"For $k \neq 1$, this is greater than the combined time intervals $2T$ recorded between events OP and PQ by B and C. But should not the time lapse between two events agree? This is one form of the so-called clock paradox." (In Fig 7.6 B = S', C = S'').

How can you think like that? T, the time laps in S' between the events O and P in the model, is not the real time that the clock in S' shows, but it is **the mathematical time from the model**.
If you make a model in which you convert physical quantities, then you must take these transformations into account all the way!
In the k-calculus model we have $c = 1$, $v = 1/10{,}000$ (e.g. Earth's velocity around the Sun) and e.g.

***t* = *1 second* in S becomes**
$t' = (1 + v^2)^{1/2}$ *seconds* in S'.

To compare quantities from reality, one must convert these quantities from the mathematical model to the real model, to reality!

In the book [7.1], we continue on to section 2.12 The Lorentz transformations.
This section is also nonsense because you base your calculations on incorrect grounds.

The error originates in the assumption that the time in the reference system in motion, t', is of the expression
$t' = kt$, or $Q = (k + k^{-1})t'$ there
$k = ((1 + v) / (1 - v))^{1/2}$ and t is the time in stationary reference system!

According to Fig. 7.5 and Fig. 7.6, the relationship between the time intervals in S and S' is as follows:

$$t' = t(1 + v^2)^{1/2}$$

This is a relation between time laps in S and S' that results from the k-calculus model!

7.2.12 The Lorentz transformations

"We have derived a number of important results in

special relativity".
What results? If you got any results then they contradict the mathematical model you created!

From Fig. 2.17, according to the author of [7.1], the following relations result:

$$t' - x' = k(t - x),\ t + x = k\,(t' + x') \qquad (2.7)$$

We have seen before that the relation $t' - x' = k(y - x)$ is wrong. According to calculations in 2.18 above, this relationship would look like this:

$$t' - x' = (t - x)(1 + v^2)^{1/2}$$

and then $t' - x' = k(y - x)$ can take place only if $v = 0$.

Furthermore, they come to

$$t' = (t - vx)/(1 - v^2)^{1/2},\ x' = (x - vt)/(1 - v^2)^{1/2} \quad (2.8)$$

and these relations are called **special Lorentz transformation** (in this model).
In chapter **3 The key attributes of special relativity** they make further calculations and then they come to the usual LT:

$$t' = (t - vx/c^2)\gamma\ ,\ x' = (x - vt)\gamma \qquad (3.12)$$

where $\gamma = 1/(1 - v^2/c^2)^{1/2}$ is called the Lorentz factor.
But I have to ask another question here:

If the formula (2.8) represents Lorentz transformations from k-calculus, the model we have analyzed, and (3.12) represents Lorentz transformations from reality then we should be able to derive (3.12) from (2.8) by applying a reverse procedure than the one we did than we created the k-calculus model! Is not it like that?

Why else has the k-calculus model been built? To what use?

We compare one more time the physical quantities from the two models

- $t' = (t - vx) / (1 - v^2)^{1/2}$ from (2.8)
- $t' = (t - vx / c^2) / (1 - v^2 / c^2)^{1/2}$ from (3.12)

- $x' = (x - vt) / (1 - v^2)^{1/2}$ from (2.8)
- $x' = (x - vt) / (1 - v^2 / c^2)^{1/2}$ from (3.12)

Say we have a conversion method from (2.8) to (3.12) and one from (3.12) to (2.8).

- $c = 1$ from (3.12) to (2.8), this works
- $1 = c$ from (2.8) to (3.12), this does not work

It is easy to go from (3.12) to (2.8) by replacing c with 1 but the other way around is more responsive. Which 1 should be replaced with c and in which places?

It is not obvious!

You have created k-calculus, you have derived LT in it but you can not translate the result back to the real model? Why?

My answer is the following.
No model of reality that uses Lorentz transformations can be without contradiction!

7. Clarification

We explain once again how the model from k-calculus works, Fig. 7.7.

If we compare this model, M_k, with reality, we get the following:
(all quantities from the model are marked with an index $_k$).

$c_k = c / c = 1$; speed of light in M_k

$v_k = v / c$; the velocity in M_k at which S' moves relative to S

$t_k = t$; the only physical quantity that is the same as in reality

x_k is expressed in light-units, e.g. light-seconds if t is measured in seconds

This forms the basis of model M_k. Light signals transmitted from S to S' move in a line that is 45° to the t-axis. Based on these assumptions that we build into the M_k model, we can calculate all other elements such as distances, time intervals, etc.

Once we have decided which mathematical model to use, only mathematics applies!

We calculate the distance between S and S' at time t_k.

$$x_k = v_k t_k$$

From the model we see clearly that the time interval in S', t'_k, is greater than the same time interval t_k in S, (only in M_k). It appears from the triangle $Ot_kt'_k$.

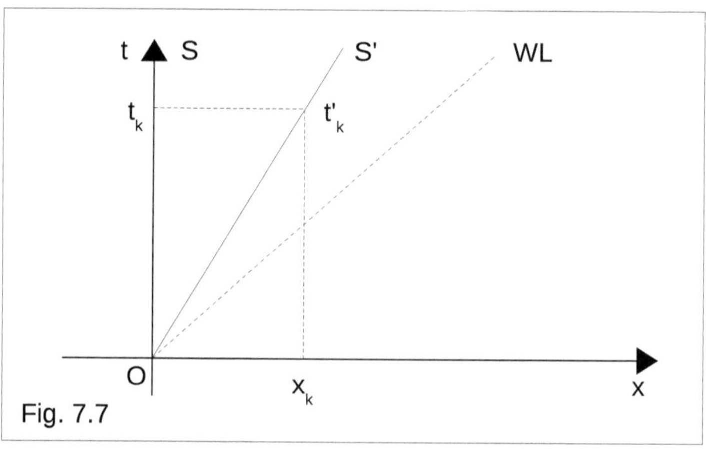

Fig. 7.7

$$t'_k = (t_k^2+v_k^2t_k^2)^{1/2} = t_k(1+v_k^2)^{1/2}$$

We see that there is a conversion factor between time intervals in S and S'. But that is not the factor mentioned in the book [7.1]. Conversion factor is

$q = (1+v_k^2)^{1/2}$ in the mathematical model M_k

$k = ((1 + v) / (1 - v))^{1/2}$ in the book [7.1]

It is because of this factor k that the contradiction arises in the model!

7. Conclusion

If we build our mathematical model correctly, if we apply mathematics, physics and logic correctly, then we shall not come to any contradiction!

Similar analysis of different concepts within SR is also done in [7.3-7.4].

7. References

[7.1] *Introducing Einstein's Relativity;* Ray d'Inverno, Chapter 2; 1992

[7.2] *Light - the absolute reference in the universe*; fourth edition; Jan Slowak; 2022

[7.3] *Special Relativity is Nonsense*; third edition; Jan Slowak; 2020

[7.4] Physics Essays: *Mathematics shows that the Lorentz transformations are not self-consistent*; Jan Slowak; 2020

8. Experiment that support SR?

Researchers who claim that there is an experiment that supports / confirms the special theory of relativity directly insults the millions of students who study mathematics at a highschool level. Why? Because the arguments raised during these experiments, the argument that is said to confirm SR go against the current mathematics and logic.

At the link [8.1] you will find a course on the special theory of relativity. The first page says:
Special Relativity with Brian Greene, Physicist, Author, Co-founder, World Science Festival; Professor of Physics and Mathematics at Columbia University

Einstein's Special Relativity upended our understanding of space, time and energy. While the ideas are subtle, they only require high school algebra, so join this math-based introduction with acclaimed physicist and author, Brian Greene.

I have taken the course and received my certificate. I attach it at the end of the book. It is not proof that I master the special theory of relativity 100%. As Brian Green says, and not just him, it only takes mathematics at the high school level to be able to understand and work with SR.

I read the course as a result of an exchange of letters between me and Ulf Gran from the University of Gothenburg. He assumed that I needed to take a basic course on the special theory of relativity.

In the course on [8.1], Brian Green claims that one of the experiments that proves the time dilation is The Hafele – Keating experiment.

I quote from Wikipedia, see link [8.2]:
The Hafele–Keating experiment was a test of the theory of relativity. In 1971, Joseph C. Hafele, a physicist, and Richard E. Keating, an astronomer, took four cesium-beam atomic clocks aboard commercial airliners. They flew twice around the world, first eastward, then westward, and compared the clocks against others that remained at the United States Naval Observatory. When reunited, the three sets of clocks were found to disagree with one another, and their differences were consistent with the predictions of special and general relativity.

I argue that this experiment can not be used as evidence to verify the special theory of relativity, SR.

SR is based on the following:

All thought experiments with the help of which Lorentz transformations are derived are based on two inertial reference systems. These are reference systems that move in line at a constant speed relative to each other.

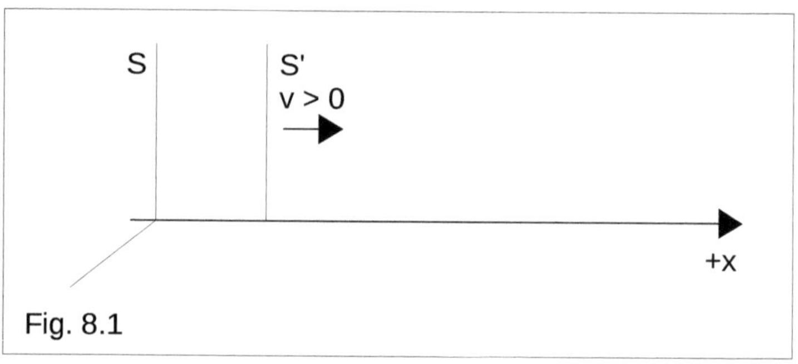

Fig. 8.1

The Hafele – Keating experiment involves the following two reference systems:

1) A point on Earth where control clocks are located. This point moves around the Earth's axis in a circle, the Earth moves around the Sun in an elliptical orbit. Where somewhere do you see the straight line on which an inertial reference system from SR would move?

2) An aircraft, as a reference system in motion. The aircraft moves around the Earth in a circular orbit. Can anyone claim that the aircraft is an inertial reference system?

The two reference systems from SR move at a constant

speed relative to each other. This means that the relative speed, $v > 0$, the speed between the point on Earth where reference clocks were located and the aircraft, must have a constant value.
Can anyone claim that the speed between the point on Earth and the aircraft was constant?
Fig. 8.2.

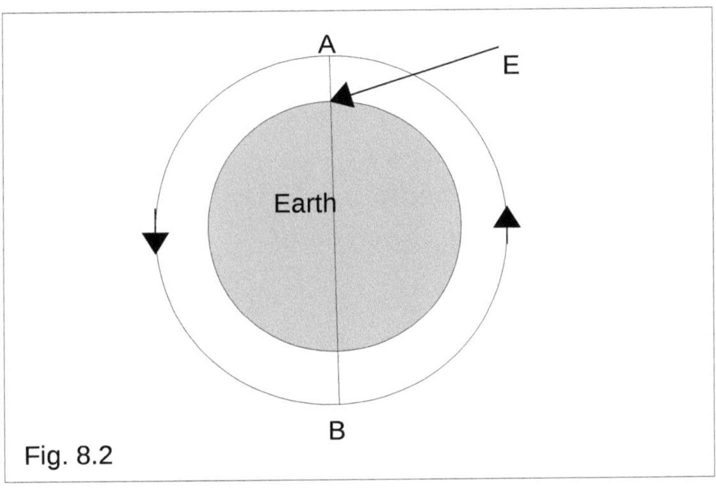

Fig. 8.2

We say that the aircraft takes off from an airport and when it passes point E it has reached its constant speed v. Relative to which reference system is the speed of the aircraft considered to be constant? I would argue that if the speed of the airplane is constant in its orbit around the Earth then it would be constant relative to the point where reference clocks are located if this point was in the center of the circle.

The relative speed between the point on Earth where the reference clocks were and the aircraft was not constant during this experiment!
Who can claim that this experiment falls within the framework of the model of the special theory of relativity?

References

[8.1] *https://worldscienceu.com/courses/special-relativity-world-science-u/*
[8.2] *https://en.wikipedia.org/wiki/Hafele%E2%80%93Keating_experiment*

9. Horizontal light clock

In my book [9.1], chapter *Analysis of time dilation*, I go through the experiment with a **vertical light clock**. There I show that in all literature the model with vertical light clock is shown incorrectly. **There is no explanation for the authors' claim that the light inside the vertical light clock moves obliquely**. It's wrong. In that experiment, the light goes vertically and nothing else!

We are now going through the experiment with a **horizontal light clock**. A horizontal light clock is a device consisting of two parallel vertical mirrors. This light clock moves in the same direction as the light signal used in the experiment. The distance between the mirrors is L.

We will analyze three thought experiments with a light clock.
The light signal we use in the experiment is sent from left to right in our figures.

1) Horizontal light clock is in absolute rest in space

The time the light signal needs to pass from one mirror to the other and back is $t_0 = 2L / c$,

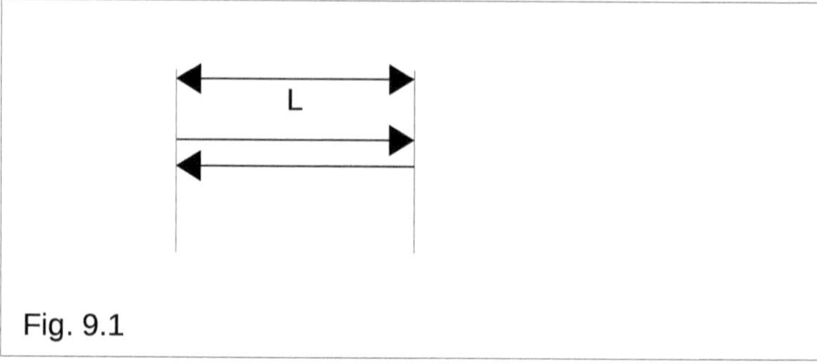

Fig. 9.1

where L is the distance between the mirrors and c is the speed of light. This is the same time you get in a vertical light clock with the same dimensions as our horizontal light clock and which is in absolute rest in space.

2) Horizontal light clock is moving from left to right at constant speed v

While the light signal goes from the left mirror to the right, the whole device has time to pass a small distance, Fig. 9.2.

$$ct_1 = L + vt_1 \rightarrow$$
$$L = t_1(c - v) \rightarrow$$
$$t_1 = L / (c - v)$$

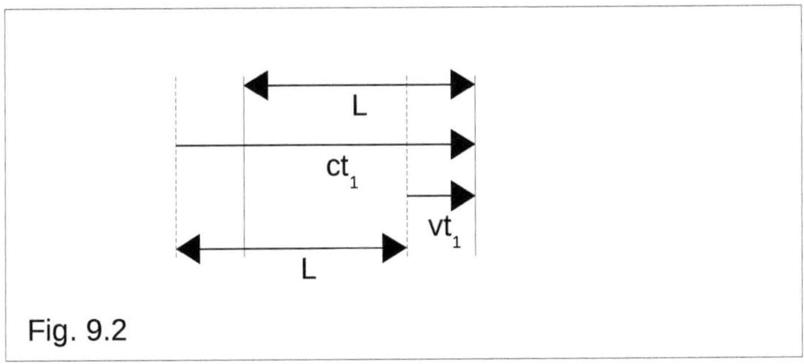

Fig. 9.2

Now we look at the part of the experiment when the light signal is reflected and go back to the first mirror, Fig. 9.3.

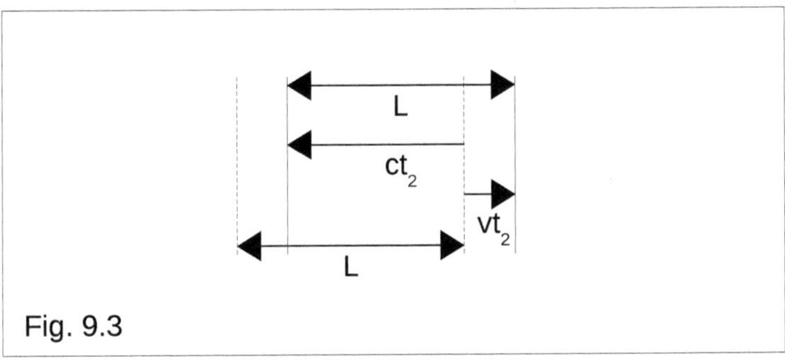

Fig. 9.3

$ct_2 = L - vt_2 \rightarrow$
$L = t_2(c + v) \rightarrow$
$t_2 = L/(c + v)$

The total time that the light signal needs to go from one mirror to the other and back is the sum of t_1 and t_2:

$$tick\text{-}tock = t_1 + t_2 \rightarrow$$
$$tick\text{-}tock = L / (c - v) + L / (c + v) \rightarrow$$
$$tick\text{-}tock = 2Lc / (c^2 - v^2)$$

We can express this time using t_0, tick-tock time for a light clock in absolute rest.

$$tick\text{-}tock = (2L / c)c^2 / (c^2 - v^2) \rightarrow$$
$$tick\text{-}tock = t_0 c^2 / (c^2 - v^2) \rightarrow$$

The factor $c^2 / (c^2 - v^2)$ is always ≥ 1.

From the *tick-tock* formula, you can calculate the speed of the horizontal light clock in space. You calculate t_0, you measure t and then you can calculate the absolute speed in the direction you sent the light beam in the horizontal light clock

$$v = c(1 - t_0 / t)^{1/2}$$

3) Horizontal light clock is moving from right to left at constant speed v

We get the same result as in 2).

9. References

[9.1] Slowak J 2020 *Special Relativity Is Nonsense*

10. Doppler effect/ Doppler shift

A wave (wave motion) in a medium has two characteristic properties:

wavelength λ and
frequency ν (we will use the letter f).

They are linked by the following formula:

$$\lambda f = v_m$$

where v_m is the speed of wave motion in the medium. We will talk about electromagnetic waves, in particular light waves, light signals, therefore we will have v_m = c, the speed of light in vacuum. So for light waves we get

$$\lambda f = c$$

The Doppler effect is about the change in either the frequency or the wavelength of a wave motion due to the motion of the source and/or the motion of the observer in the medium. The movements of the source and the observer can be merged into so-called relative motion. In what follows, we will deal with three different cases: the source and the observer are at absolute rest in space (in the direction in which we will make calculations), the source and the observer approach each other, the source and the observer move away from each other.

We will make the calculation of the change in wavelength due to the relative velocity between the source and the observer. This change is called Doppler shift.

1) The source and the observer are at absolute rest in space

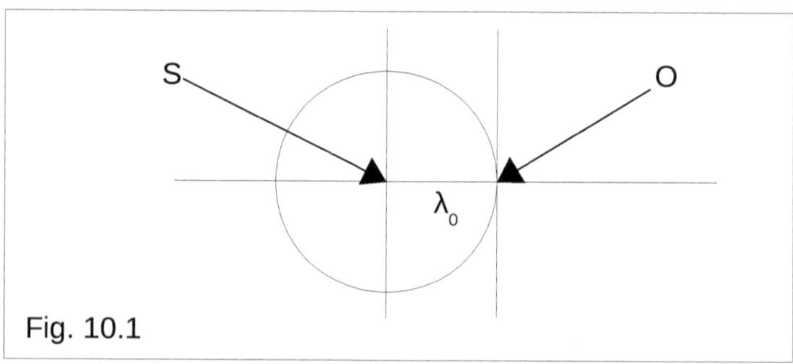

Fig. 10.1

The source sends a light signal that has the wavelength λ_0. We assume in these experiments that the distance between the light source and the observer is equal to λ_0, the wavelength that is characteristic of the wave motion in the respective medium in which we do our experiment. The time that the light wave needs to pass the distance λ_0 is

$$t_0 = \lambda_0 / c$$

We show a simplified picture: the source, the observer

and the distance between them, Fig. 10.2.

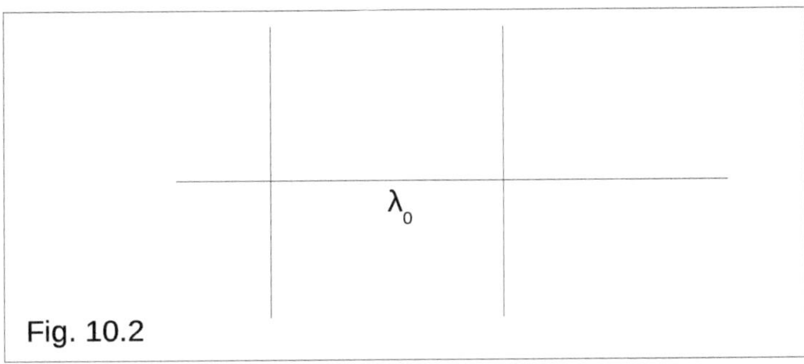

Fig. 10.2

2) The source and the observer approach each other at speed v

As the light signal reaches the observer, the light source will move a short distance, closer to the observer.

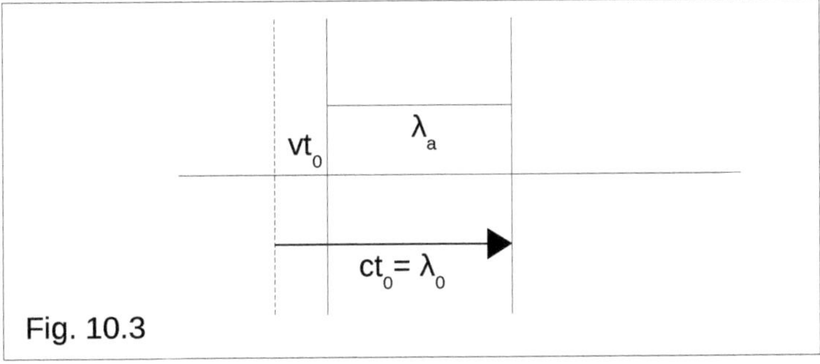

Fig. 10.3

From the Fig. 10.3 we get the relation

$ct_0 = vt_0 + \lambda_a \rightarrow$
$\lambda_a = ct_0 - vt_0 \rightarrow$
$\lambda_a = (c - v)t_0 \rightarrow$
$t_0 = \lambda_a / (c - v) \rightarrow$
$\lambda_0 / c = \lambda_a / (c - v) \rightarrow$
$\boldsymbol{\lambda_a = \lambda_0(c - v)/c}$

3) The source and the observer move away from each other at speed v

When the light signal reaches the observer, the light source will move a short distance away from the observer.

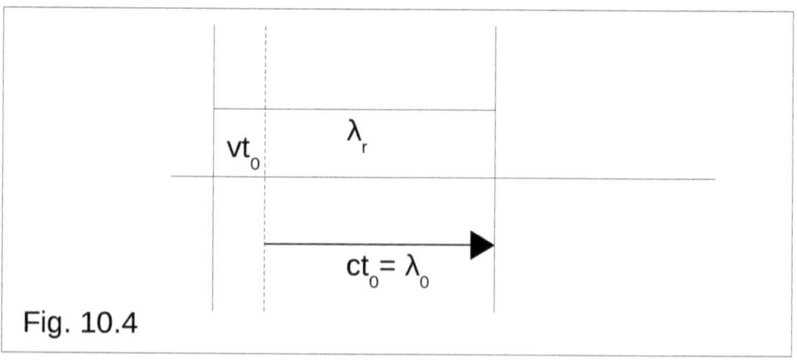

Fig. 10.4

From the Fig. 10.4 we get the relation

$\lambda_r = vt_0 + ct_0 \rightarrow$
$\lambda_r = (c + v)t_0 \rightarrow$
$t_0 = \lambda_r / (c + v) \rightarrow$
$\lambda_0 / c = \lambda_r / (c + v) \rightarrow$
$\boldsymbol{\lambda_r = \lambda_0(c + v)/c}$

The change from λ_0 to the measured wavelength λ (λ_a or λ_r) is called the Doppler shift, z, [10.1].

$$z = (\lambda - \lambda_0)/\lambda_0 \rightarrow$$

$$\boldsymbol{z + 1 = \lambda/\lambda_0}$$

If $\lambda > \lambda_0$, $\lambda = \lambda_r$, we are talking about redshift , otherwise, $\lambda = \lambda_a$, about blueshift. From the above formulas we get:

$z_r = v / c$ for redshift

$z_a = -v / c$ for blueshift

These formulas depend only on the speed of light and on the relative speed between the light source and the observer.
I think there is only the classic Doppler shift. See my book [10.2].

10. References

[10.1] *https: / / en.wikipedia.org / wiki / Redshift*
[10.2] Slowak J 2015 *Redshift factor, Absolute redshift, Galaxies red / blue distribution*

11. The pole in the barn paradox

There are many paradoxes within the special theory of relativity and one tries to convince everyone that it is not actually about any paradoxes. Feel free to check out the following section from the link *https://worldscienceu.com/lessons/32-5-the-pole-in-the-barn-paradox/*

Short description:
- pole: 15 feet (at rest)
- barn: 10 feet (at rest)

At rest they do not fit.

Scenario: pole moves toward barn with $v = (12/13)c$, where c is the speed of light. Lorentz factor is calculated to $\gamma = 13/5$.

Barn's perspective: pole is Lorentz contracted,
pole's length = $15/\gamma = 5.8$ feet
→ fits

Pole's perspective: barn is Lorentz contracted,
barn's length = $10/\gamma = 3.8$ feet
→ doesn't fit

One reality, two different answers! It's just absurd!

Imagine a critical situation where it would be about life and death. A doctor will operate and ask the question to two observers. After which of the two should he make his decision?

So what does it matter that one observer perceives reality in one way or another and the other observer perceives reality in a different way?
Both should come to the same conclusion because we only have one reality and it is as it is!

12. Summa summarum

In 1905, Einstein wrote an article that became the start of a new theory: The Special Theory of Relativity. This year, 2023, it has passed

118 years

since this article was published. But what is interesting is that after all these years, there are still researchers who doubt this theory, who counter-argue it.
How can that be?

I consider myself an independent researcher and have analyzed many of the concepts covered in this theory. I can say that this theory is fundamentally wrong, it is nonsense!

I ask some questions to researchers who defend SR:

Reflection 1:
In the book [12.4], chapter Analysis of time dilation, pages 26-33, I explain inaccuracies regarding the horizontal light clock.

Question 1:
Why does the light beam in the vertical light clock that is in motion move obliquely?

Reflection 2:
In [12.6] the derivation of Lorentz Transformations is done. I analyze this derivation in [12.4], chapter Derivation of Lorentz transformations, Example 2, pages 36-38.
Special cases 1 and 2 are shown there:

SC1) $x' = 0, x = vt$

SC2) $x = 0, x' = -vt'$

Question 2:
Why not

SC1) $x' = 0, \boldsymbol{x = vt'}$

SC2) $x = 0, \boldsymbol{x' = -vt}$?

If you do this, you will come to Galileo's transformations without having to use the light postulates.

Reflection 3:
In his derivation of Lorentz's transformations in [12.7], Einstein makes the following assumption:

$x = ct, x \geq 0$

and

$x = -ct, x \leq 0$

and then he use these two equations in one and the same system of equations.
I analyze this derivation in [12.4], chapter Derivation of Lorentz transformations, Example 3, pages 38-41.

Question 3:
How can today's scientists turn a blind eye to such a mathematical error?

Reflection 4:
In my book [12.4], chapter Michelson-Morley experiment, 1887, pages 51-70, I make my own analysis of this experiment. I use the same physics, mathematics and logic as in all my work on SR and I come to the same result as the Michelson-Morley experiment from 1887. My theoretical analysis of the experiment perfectly reflects the results from the real experiment from 1887.

Question 4:
How is it possible that I get the same result?

Reflection 5:
In my book [12.4], chapter Relativity with classical physics, pages 83-89, I make a comparison between the Lorentz factor and other four factors that I call relativity factors. For the example of the Earth's motion around the Sun, $v = 30\ km/s$, there is the difference between $\pm\ 0.0001$.

Question 5:
Was the precision in the Michelson-Morley experiment

from 1887 better than ± *0.0001*?

Reflection 6:

In my book [12.4], chapter SR and spacetime, pages 107-112, I refer to the book [12.8]. This book uses the so-called modified Pythagorean theorem.

(Hypotenuse)^2 =
(Longest catheter) ^{2} – *(Shortest catheter)* ^{2}

The longest catheter has as unit *year,* unit of time! The shortest catheter has as unit *light year*, unit of length!

Question 6:

How can professors, researchers teach a theory that uses such trivial errors?

Also follow my work in my published articles [12.1-3] and books [12.4-5].

12. References

[12.1] Slowak J 2020 *Physics Essays: Mathematics Shows That the Lorentz Transformations Are Not Self-Consistent*

[12.2] Slowak J 2020 *SCIREA J. Phys.: Lorentz Transformations and Time Dilation Do Not Verify Reality*

[12.3] Slowak J 2020 *SCIREA J. Phys.: Lorentz Transformations - The Sound Versus The Light*
[12.4] Slowak J 2020 *Special Relativity Is Nonsense*
[12.5] Slowak J 2022 *Light – The Absolute Reference in the Universe*
[12.6] Harris R 2008 *Special Relativity in Modern Physics*, chap. 2
[12.7] Einstein A 2006 *Den speciella och den allmänna relativitetsteorin*; Första delen; Om den speciella relativitetsteorin; Swedish
[12.8] Holst S, *Rumtid – en introduktion till Einsteins relativitetsteori*, 2006, Swedish

13. My last derivation of Lorentz transformations

We follow similar thinking as in [13.1], pages 14-15. We consider two inertial reference systems S and S'. Their x-, x'-axes coincide and we will only look at events that occur on their common x-, x'-axes. At the beginning of each thought experiment, S and S' are at the same point. S' moves to the right at constant speed $v > 0$.

We call our experiments special cases.
For each experiment we make a mathematical model. After this model has been developed, we make calculations on the model by applying current mathematics and logic.

An event is denoted by $E = (x, t)$ for S and $E' = (x', t')$ for S'. Remember that we are only dealing with one event at a time but we are referring to it from two different frames of reference.

It is assumed that the relationships between the coordinates in the two reference systems are linear. See the next system of equations consisting of the equations LEx' and LEt'.

LEx': $x' = Ax + Bt$
LEt': $t' = Cx + Dt$

where A, B, C, D are constants.

Special case 1:
After S' has passed S and has moved to the right, an event occurs in S'-origin. See the model in Fig. 13.1.

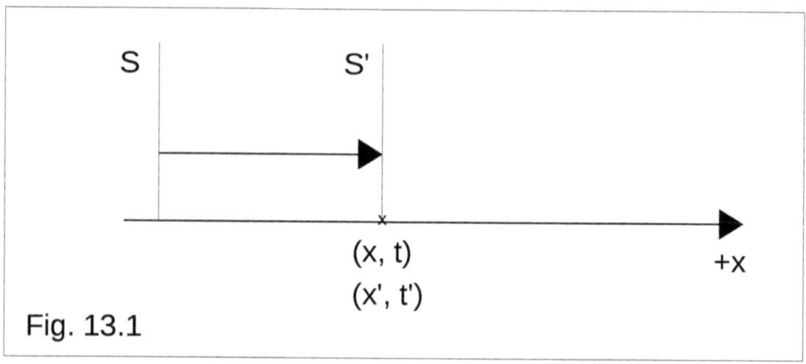

Fig. 13.1

Our attention is directed to the reference system S'. When the event occurs, the time $\boldsymbol{t'}$ is recorded.
We now determine the value of our four coordinates: x, t, x', t'.

We have $\boldsymbol{x' = 0}$, the event occurred in S'-origin.
The distance between S and S' is $\boldsymbol{x}$.
x is positive, it is to the right of the S-origin.

During t', S' moved at velocity $v > 0$ by a distance x.

$\rightarrow t' = x/v \rightarrow x = vt'$.

Both t and t' are positive, $t > 0, t' > 0$. $\rightarrow$

SC1x': $x' = 0$
SC1x: $x = vt'$

Special case 2:
After S' has passed S and has moved to the right, an event occurs in S-origin. See the model in Fig. 13.2.

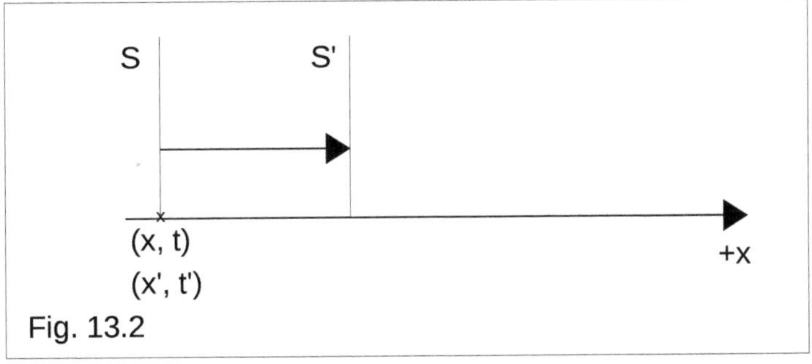

Fig. 13.2

Our attention is directed to the reference system S. When the event occurs, the time $\boldsymbol{t}$ is recorded. We now determine the value of our four coordinates: x, t, x', t'.

We have $\boldsymbol{x = 0}$, the event occurred in S-origin. The distance between S and S' is $\boldsymbol{x'}$. x' is negative because it is to the left of the S'-origin. During time t, S' moved at velocity $v > 0$ by a distance $|x'|$.

$\rightarrow t = |x'|/v \rightarrow x' = -vt.$

Both t and t' are positive, $t > 0, t' > 0. \rightarrow$

SC2x: $x = 0$
SC2x': $x' = -vt$

We replace SC2x and SC2x' in LEx' and LEt' and make calculations (* is for multiplication):

$-vt = A*0 + Bt$

$t' = C*0 + Dt$

$\rightarrow$

$\boldsymbol{B = -v}$ and $\boldsymbol{t' = Dt}$

We replace SC1x' and SC1x in LEx' and LEt' and make calculations:

$0 = Avt' + Bt$

$t' = Cvt' + Dt$

From $0 = Avt' + Bt$ and $B = -v \rightarrow Avt' - vt = 0 \rightarrow$

$Avt' = vt \rightarrow$

$At' = t \rightarrow \boldsymbol{t = At'}$

In a linear relationship between two variables, their ratio is always the same.
From $t'/t = 1/A$ and $t'/t = D \rightarrow 1/A = D \rightarrow$

$AD = 1 \rightarrow$

$\boldsymbol{A = 1/D \rightarrow D = 1/A}$

From $t' = Cvt' + Dt$ and $t' = Dt \rightarrow Cvt' = 0 \rightarrow \boldsymbol{C = 0} \rightarrow$
Now our linear equations have become

LEx': $x' = Ax - vt$

LEt': $t' = (1/A)t$

We now use the symmetry between S and S' by changing x to x', t to t', x' to x, t' to t and v to $-v \rightarrow$
(ILE for inverse linear equations)

ILEx: $x = Ax' + vt'$

ILEt: $t = (1/A)t'$

Both equation systems LE and ILE apply at the same time!
From $t' = (1/A)t$ and $t = (1/A)t' \rightarrow$ by addition $\rightarrow$
But $t > 0, t' > 0 \rightarrow t + t' > 0 \rightarrow$

$t' + t = (1/A)(t' + t) \rightarrow$

$(1/A) = 1 \rightarrow \boldsymbol{A = 1} \rightarrow \boldsymbol{D = 1}$

$\rightarrow$

LEx': $x' = x - vt$

LEt': $t' = t$

and

ILEx: $x = x' + vt$

ILEt: $t = t'$

Within SR, these equations are referred to as Lorentz transformations.

But alas, with only two thought experiments and without mixing in motion of light, we come to the classic Galileo's transformations!
Note that in SC1 and SC2 we have not used the light speed.
It was not needed.
How can we arrive at different results in the same thought experiment?

In my derivation I have in SC1 arrived at

SC1x: $x = vt'$

while in [13.1] to

SC1x: $x = vt$

and in my derivation I have in SC2 arrived at

SC2x': $x' = -vt$

while in [13.1] to

SC2x': $x' = -vt'$

But look at the model in Fig. 13.1. The distance between S-origin and S'-origin when the event occurs is the same for S and S'. We only have one stretch SS'.

→

$$d(SS') = \boldsymbol{vt'} = \boldsymbol{vt} \rightarrow \boldsymbol{t} = \boldsymbol{t'}.$$

The same applies to the model in Fig. 13.2. and then we clearly see that we can use either vt or vt', they are the same!

If you use $t = t'$ in LT, you come to a contradiction! No more comments!

13. References

[13.1] Harris R 2008 *Special Relativity in Modern Physics*, chap. 2

14. A few excerpts from my notebooks

2020-02-14

If you sit in a bus and look at distant objects, you see that they are moving. But do they move in reality in the same way as we perceive them? No!
It's just an illusion!

If we want to measure an object at a distance, we can only do so if we know the distance to the object.

It is said in SR that clocks in motion run slower than clocks at rest.
How does this fit with the postulate that all inertial frames of reference are equivalent?

It is said that S' moves at a constant speed relative to S which is at rest. Then the clock runs slower in S' than in S.
At the same time, S' can imagine that it is its reference system that is at rest and S is moving away with speed v. Then the clock in S runs slower than that in S'.
These two reference systems are symmetrical, there is no difference between them.

If S' specifies the formula for time dilation, then $t' = t\gamma$.
If S specifies this formula, it becomes $t = t'\gamma$.

From here you get that $\gamma = 1 \rightarrow v = 0$!

2020-02-29

Mixing up (putting in the same equation) motion of a reference system moving at speed v, with the motion of a light signal moving at speed c, as done in [14.1], is extremely strange.

Compare the next two models, the first showing the equation for an object moving at speed v and the second showing the equation for a light signal moving at the speed of light c.

Try putting these two equations, $x = vt$ and $x = ct$, into one and the same general equation! What do you get?

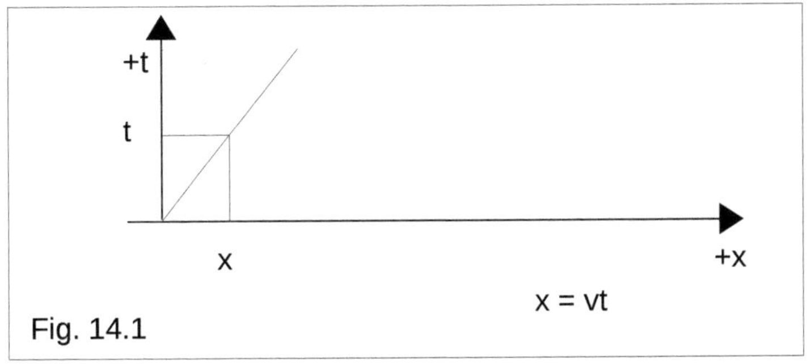

Fig. 14.1

$Ax + Bt + C = 0$

$x = vt$

$\rightarrow$

$$Avt + Bt + C = 0$$

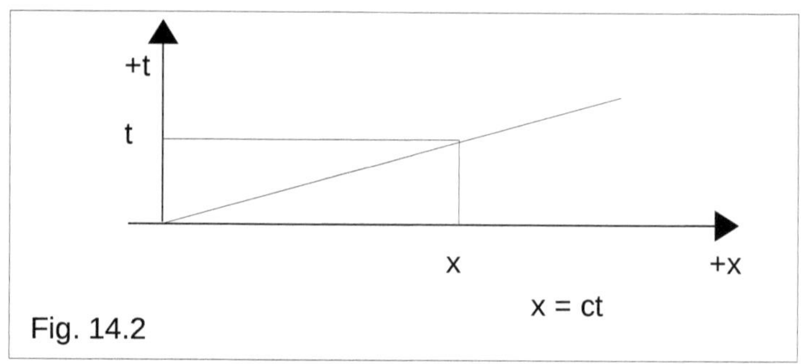

Fig. 14.2

$$Ax + Bt + C = 0$$
$$x = ct$$
$$\rightarrow$$
$$Act + Bt + C = 0$$

From

$$Avt + Bt + C = 0$$
$$Act + Bt + C = 0$$
$$\rightarrow v = c$$

This is absurd! Putting the motion of a Tesla car and the motion of a snail into the same system equations!

2020-04-05

You cannot connect space and time as you did in SR. The relative motion between S and S' is independent of one or more events occurring somewhere!

The relative motion between S and S' is independent of one or more light signals moving somewhere in the space!

2020-04-06

A system of two linear equations with two variables, $f(x, y) = 0$ and $g(x, y) = 0$, has a single solution: the point at which the two lines intersect.

A system of two linear equations with three variables, $f(x, y, z) = 0$ and $g(x, y, z) = 0$, has a single solution: the line at which the two plane intersect.

In [14.1], one starts from two linear equations:

LEx': $x' = Ax + Bt$,

LEt': $t' = Cx + Dt$,

where A, B, C, and D are constants.

The equation LEx' passes through the point $(x, t, x') = (0, 0, 0)$.
The equation LEt' passes through the point $(x, t, t') = (0, 0, 0)$.
The two reference systems have the x-axis and the x'-axis in common.
The plane of all points verifying LEx' passes through the point $(x, t, x') = (0, 0, 0)$.

The plane with all points verifying LEt' passes through the point $(x, t, t') = (0, 0, 0)$.

Which line is it where these two planes cross each other?

2020-04-10

In the literature regarding SR, it is usually said that two objects move relative to each other with a constant speed v.
It also describes how the light signal moves in one or the other reference system.

But in my opinion, one should describe how an object moves relative to a light signal.
As I describe in the book [14.2].

2020-12-28

We refer again to [14.1].
Lorentz transformations are described there:

LTx': $x' = (x - vt)\gamma$,

LTt': $t' = (t - vx/c^2)\gamma$,

where $\gamma = 1/(1 - v^2/c^2)^{1/2}$ is called the Lorentz factor.

These equations have been derived from

LEx': $x' = Ax + Bt$,

LEt': $t' = Cx + Dt$,

by applying three thought experiments.

We repeat these experiments with a slight modification.

SC1:
The experiment begins when the two reference systems are at the same point. Then we have
$x' = 0, t' = 0, x = 0, t = 0.$
S' moves to the right with constant speed $v > 0$. When S' reaches the point $x_d = d$, an event occurs at the S'-origin.
The event receives the following coordinates:

$$x' = 0, t' > 0$$

We replace these in LTx' and LTt':

LTx': $0 = (x - vt)\gamma$
LTt': $t' = (t - vx/c^2)\gamma$
$\rightarrow$
$x = vt = d \rightarrow \boldsymbol{t = d/v}$

We replace this in LTt':

$t' = (d/v - vd/c^2)\gamma \rightarrow$
$\boldsymbol{t' = d/v\gamma}$

SC2:
The experiment begins when the two reference systems are at the same point. Then we have

$x' = 0, t' = 0, x = 0, t = 0.$
S' moves to the right with constant speed $v > 0$.
When S' reaches the point $x_d = d$, an event occurs at the S-origin.
The event receives the following coordinates:

$x = 0, t > 0$

We replace these in LTx' and LTt':

LTx': $x' = (0 - vt)\gamma$
LTt': $t' = (t - 0)\gamma$
$\rightarrow$
$x' = -vt\gamma = -d \rightarrow$
$\boldsymbol{t = d/v\gamma}$
$t' = t\gamma \rightarrow \boldsymbol{t' = d/v}$

The way we performed the two thought experiments means that t from SC1 is equal to t from SC2

$t = d/v$ and $t = d/v\gamma \rightarrow$
$\gamma = 1 \rightarrow$
$v = 0$

In the same way, the times t' must be equal in the two thought experiments:

$t' = d/v\gamma$ and $t' = d/v \rightarrow$
$\gamma = 1 \rightarrow$
$v = 0$

How so? Can't one do two thought experiments in which the distance between S and S' could be d? Isn't that ridiculous?

2021-01-05
Thoughts on the Big Bang and black holes.

2021-03-10
What is an event?
An event is something that happens, for example that a short light signal occurs in a reference system. Such a thing can NOT affect anything in the reference system or in an observer or in another reference system or an observer in it.

How can one say that if two observers measure distance and record the time of this event that this would affect the passage of time or anything else?

How can one say that numerous experiments verify SR when SR itself has no basis in the reality in which these experiments are conducted?

2021-01-05
Thoughts on the Big Bang and background radiation.

2022-05-05

What happens if you derive two equations and to determine the constants you use two conditions such as

$x = vt$ and

$x = ct$?

No matter how it is done, these two conditions must be reflected in the two equations.
It goes without saying that the variable x cannot vary both as vt and ct in one and the same system of equations!
Example:

LE1: $ax + by + c = 0$

SC1: $x = cy$

SC2: $x = vy$

$\rightarrow$

$v = c$

14. References

[14.1] *Special Relativity in Modern Physics*, chap. 2, Harris R 2008
[14.2] *Light - the absolute reference in the universe*; fourth edition; Jan Slowak; 2022

15. My correspondence with 'established researchers'

I have written to many researchers, physics departments, asked them some questions, asked them to review my research. Most people don't answer!

Why?
Don't they have time to answer any questions? Don't they have some obligation to answer?
Could it be that they don't know how to answer?
Could it be that they don't master SR very well and they don't dare to make mistakes?

Could it be that they know that SR is wrong but don't dare to talk about it in order not to lose their job?

Below are some cases:

Peter Olofsson
peter.olofsson@ju.se
Jönköping University

I have written about 70 letters to Peter Olofsson. I have received answers to most of them. But whether what he explains is right or not remains to be seen.

The conversation with Peter Olofsson was in Swedish. What I reproduce here is but my own translation.

My letter No. 32:
Peter Olsson:
Mathematically, it seems correct, in that vt' + ct' in the upper figure is as long as ct in the lower one, as you have drawn it.

Me:
I consider them equal. It is the same distance that we treat in the model, the distance between the S origin and the point E/E'. We also only have one light signal, which moves this stretch. I can't imagine it being any other way.

Peter Olsson:
If there is something physically wrong, I don't know. Feels like you somehow assume that S and S' measure the same, which I don't think you can do. I'm unsure about the physics and leave it unanswered.

Me:
If you are unsure about the physics, as you have said before, then it will be difficult to continue.
...
What I would like is the following:
Make a similar mathematical model of what is

included in the derivation of LT in [15.1], [15.2]. The best would be if you make a drawing like I did. If that doesn't work, I would settle for a description of the model.

We have two reference systems S, S' and an event $E = (x, t)$, $E' = (x', t')$:
What does the x coordinate mean for S? How does it relate to E?
What does the t coordinate mean for S? How does it relate to E?
What does the x' coordinate mean for S'? How does it relate to E'?
What does the t' coordinate mean for S'? How does it relate to E'?

I draw the basis of the thought experiment covered in SR, LT.

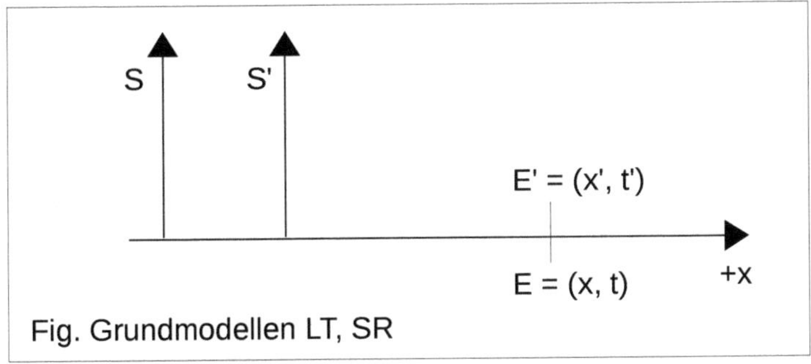

Fig. Grundmodellen LT, SR

We know that the thought experiment begins when the two reference systems are at the same point and then their clocks are also reset.
The 4 questions above can perhaps be asked in a different way. I'm mostly looking for what t, t' refers to. But the meanings of x, x' are also important.
For t, t' are times of S, S' clocks. How is the relationship between S and E, how is the relationship between S' and E'.

Without these regulations, no calculations can be made!

My letter No. 34:
Me:
You have never answered my concrete questions, e.g.:
Shouldn't one get an equality when substituting a solution into original equations?
This particular question was a purely mathematical question.

My letter No. 35:
Peter Olsson:
Yes, I can't get into questions related to physics. Haven't dealt with it since high school and that was a while ago.

"Shouldn't you get an equality when you substitute a

solution into original equations?"
Yes, you should get that.

Me:
This should also be the case when verifying LT.
But no!

LT1, SC1 → $0 = 0$
LT2, SC1 → $t' = t / \gamma$
LT1, SC2 → $t' = t\gamma$
LT2, SC2 → $t' = t\gamma$
LT1, SC3 → $t' = t(1-v / c)\gamma$
LT2, SC3 → $t' = t(1-v / c)\gamma$

It is only one of six variants that gives equality!

Conclusion: LT is not self-consistent! Purely mathematical!
So you don't need a physicist to look at this!

My letter No. 39:
Peter Olsson:
I don't really know what you are asking about, what you mean by "the distance between S' and S". Isn't that just x - x'? If you now have LT in mind, the question is what coordinates one and the same point has partly in S', and partly in S. In your figure you see e.g. that the origin in S' is the same as x = vt in S, and that the origin in S is the same as x'=-vt' in S'.

My letter No. 40:
Peter Olsson:
Your transformation is of course correct for two fixed coordinate systems. If the origin in S′ has x-coordinate d in S, then $x = d + x'$ *applies.*

Me, now:
But then when S' has x-coordinate = $vt' \rightarrow x = vt' + x'$!
But then this does NOT verify Lorentz transformations!
See chapter 4 i this book:

LTx: $x = (x' + vt')\gamma$,

LTt: $t = (t' + vx'/c^2)\gamma$.

$\rightarrow (x' + vt')\gamma = x' + vt' \rightarrow \gamma = 1 \rightarrow v = 0$

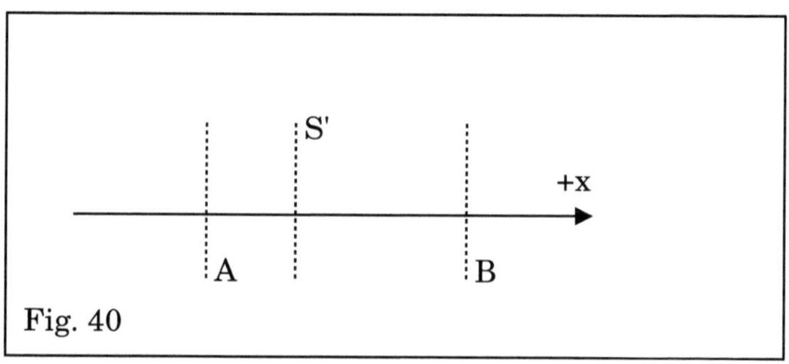

Fig. 40

The distance between A and B is x.
The distance between A and S' is d or d'
The distance between S' and B is x'.
For each moment t, S knows that the distance $d(SS')$ is

$d(x, t) = vt.$

For each moment t', S' knows that the distance $d(SS')$ is $d'(x', t') = vt'$.

Based on these assumptions, can we then write distance relations between points A, S' and B or (S; S'; E, E') as below?

$$x' = x - vt$$
$$x = x' + vt'$$

We could not agree on the relationships between the distances between this three points.

Why? I know! If Peter Olsson would have answered Yes to my question, the answer would lead to a contradiction with LT. He didn't want that!

We have since ended up in discussion about mathematical models:

My letter No. 52:
Peter Olsson:
I know what a mathematical model is, I have been dealing with them in my research for over 25 years. The problem is that you think a mathematical model is identical to a figure. As you can see, there is nothing

about figures in the quotes, it says e.g. "purely mathematical terms". The model for two coordinate systems in relative motion requires no figures, it is pure mathematics. It says that $x' = 0$ corresponds to $x = vt$, which does not need to be drawn.

Me:
In every single book where you talk about coordinate systems, you find a figure with two axes that run at right angles to each other.
...
So I don't understand what you have against figures. I don't understand what you have against mathematical models that reflect a concrete physical problem.
...
And if in ... there are figures on thought experiments why should I not be able to use figures in my proof?
...
The mathematical model we have created has no distance function. It is clear from the Cartesian coordinate system how the distance between two points is calculated. We don't need to define this.

Peter Olsson:
Of course it has a distance function. Within each coordinate system, the Euclidean distance is used.

My letter No. 53:
Peter Olsson:
I don't mind figures. I have nothing against mathematical models. I never said either one or the other. You are the one who thinks that a figure IS a mathematical model. Read the quotes you sent me. It says nothing about figures. It is about mathematics, and mathematics is axiomatic and deductive in nature. Figures are good for illustration and clarification, but their role must not be exaggerated.

Me:
It hurts you to say You look at your deceptive figure!
I previously sent you a copy of pages 14-15 of [2].
Is even the figure from Harris's book misleading?
Is the figure from Einstein's book also misleading?

Peter Olsson:
The figures themselves are not misleading. I should have worded it differently. But when you look at your figure with a fixed coordinate system and think you can draw conclusions about systems in motion, you are making an assumption that cannot be justified other than pure assumption.

Me:
I feel insulted about your way of answering my questions.

You enter with definite propositions (which are the result of the derivation of LT) and use these as constraints in the original problem.
It is completely wrong to do so!

If someone asks you to calculate the area of a triangle ABC, then you say that it is not possible, that the dimensions of the triangle are wrong. You do this as if you had calculated the area of another concrete triangle and there you had area = 5 and the area of ABC would give area = 7.

I don't know why you do that.
But my conclusion is that you have decided not to answer Yes to some of my questions that would lead to contradiction with SR.
That's what you've done all along. Mostly you have confused me.
I don't know why you do that.
You do this yourself or you have been told to do so, either by your employer or another group.

I see no other explanation. Your approach to the problem is wrong.
You have to collect data, make calculations and then see what the result is!

You go in with the result and deny everything that

doesn't match your result!
That's not how a scientist works.
Completely wrong, weird! And even if you were to explain yourself, how could anyone believe your explanation after tens of letters of denial!

Ulf Gran
ulf.gran@chalmers.se
Gothenburg University

I have written about 20 letters to Ulf Gran.

My letter No. 3:
Ulf Gran:
My tip is that you take a university course in SR, after that I think you will have an answer as to why there are no problems with SR, and that SR is definitely not nonsense.

Me:
I have read everything about SR, in almost every book from the university library.
I have bought over 20 books that I have at home, all about SR.

Then I want to say the following: if I would have worked at a university then I would not have had the opportunity to reach this result of mine. You don't have much time for your own projects. You are an opportunist! You don't want to swim against the current!

My letter No. 3:
Me:
But I have to defend my work and counter-argument at the same time.

No human being, no scientist, can claim to possess absolute knowledge. We are so different, we have different backgrounds and therefore there are differences in how we think.

Even if we were to take the same course in physics, we say, then we will always think in different ways.
I remember how it was at school and I think you will agree with me on this. We got the same books, we had the same teacher but...
we got different grades! Why?

My letter No. 5:
Me:
I explained to Ulf Gran that the derivation of LT in [15.1] leads to a contradiction:
... if you equate them $t' = t\gamma = t / \gamma \rightarrow \gamma = 1 \rightarrow v = 0$.

Ulf Gran:

I suggest that when you get a result in your own derivations that contradicts SR, or any other accepted part of physics, don't assume that all physicists were wrong but instead try to understand the well-established derivations that are available.

My letter No. 5:
Me:
Regarding your arguments for SR, I want to quote from the book
Big Bang or Let there be light light by Maria Gunther Axelsson:
"Therefore, the research is not about defending established theories, but on the contrary testing, questioning and trying to find gaps and inaccuracies in the theories."

My letter No. 9:
Me:
First of all, you must answer F1, F2 and F3 from
Brev till Chalmers - Teoretisk fysik – 6.pdf

My letter No. 10:
Me:
I wrote before that I don't want explanations with "words" but only with mathematical tools.

My question remains:

Why equality here

LT1, SC1 $\rightarrow 0 = 0$

but not here

LT1, SC2 $\rightarrow t' = t\gamma$

and not here

LT1, SC3 $\rightarrow t' = t(1\text{-}v/c)\gamma$

If we have a linear equation depicted in (x, y) as a line and a number of points "on" this line →
Will every verification give a EQUALITY!

If you cannot answer these questions yourself, you should see that someone else assesses my work.
...
I believe that your answers are not clear from the point of view of mathematics!
If you cannot answer questions F1, F2, F3 correctly from the point of view of mathematics, then your conclusion that my work contains mistakes is incorrect!

It's quite logical! The logic that we both invoke!

My letter No. 11:
Me:
We can imagine a more general case that I depict in Fig. SCDd:
A small description:

At the beginning of the thought experiment, the two reference systems are at the same point.
At that moment, on the x-axis, at a distance D, an event occurs.
My question is:
What do *(x, t)* and *(x', t')* represent when S' is at a distance d from the S origin?

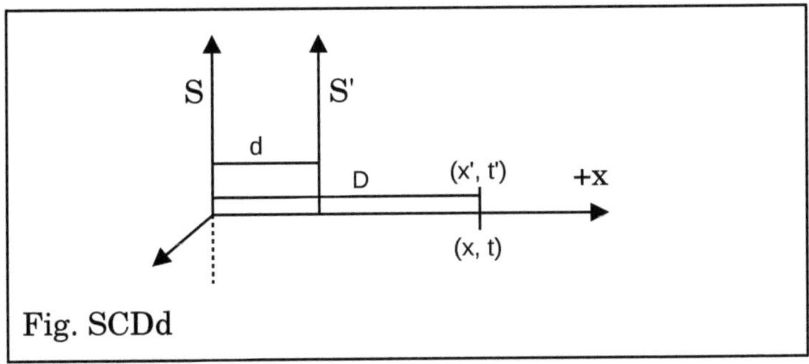

Fig. SCDd

Consider that we want to derive LT and that the above test case should help us with this.

Write the expression of the following variables in the picture Fig. SCDd.

$x =$
$t =$
$x' =$
$t' =$

My letter No. 12:

Me:
I have looked at your calculations, the ones you attached before.
These calculations, the derivation of LT, I have done dozens of times.
You, in the same way as the author of [15.1], make mistakes.

You write:
SC1: $x' = 0, x = vt, t'$ unknown.

This is wrong, that you say that t' is unknown.
It is the opposite! t' is known and t is unknown!

The event occurs in S'. S has no knowledge of this event!
Only S' can say something about the event and make calculations.
S' sees that when the event occurs then the clock shows t' !!!
It is S' that can calculate the distance to S:
This distance is vt' and nothing else!

You write:
SC2: $x' = -vt', x = 0, t$ unknown.

This is also wrong, to say that t is unknown.
It is the opposite! t is known and t' is unknown!

When the event occurs in S then $x = 0$ and the clock in S shows time t.
It is S that can calculate the distance to S': This distance is vt and nothing else!

It is because of this that I asked you to give a definition of what the coordinates of the event mean. This is the basis for all the conclusions from the derivation of LT.

I have made dozens of tens of calculations with all possible variants.

If we make calculations for SC2 as I say above we get:

SC2: $x' = -vt,\ x = 0,\ t > 0 \rightarrow$
LEx': $x' = Ax + Bt \rightarrow -vt = 0 + Bt \rightarrow -vt = Bt \rightarrow$
$\boldsymbol{B = -v}$

This contradicts the result from [15.1].

Please comment on my statements in this letter.
Please send me a definition of what the event's coordinates mean.

My letter No. 13:
Me:
It has been more than 1.5 years since we last communicated!

...

Since then, I have published three articles and a book about the special theory of relativity, SR.

My letter No. 16:
Me:
From my letter 8:
*Another clear example of a basic logical error is your question 3 in the last email, the one you call F3. There you expect to have the same relationship between T and T 'for three *different* relative movements between the inertial systems. What follows from the derivation is precisely that time and space in different inertial systems *depends* of the relative movement, and that the relationship is given by the Lorentz transformation.*

You think wrong here too:
It is not about three *different* relative movements between the inertial systems.
They are about exactly the same relative movements between the inertial systems in the 3 SC.
For two inertial reference systems S, S' which moves relative to each other at constant speed v applies to a single relationship between t and t'.
It is the formula for the time dilatation: $t' = t\gamma$. So no matter which SC we choose, the relationship $t'/t = \gamma$ is the same, constant !!!
...

Speaking of events! Check out the picture Fig. 18-1.

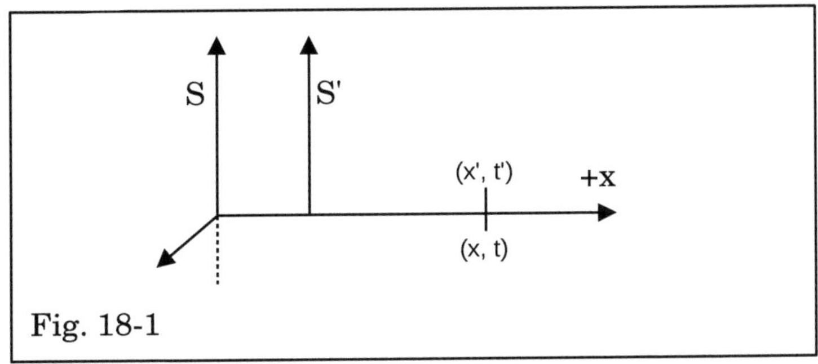

Fig. 18-1

We have two inertial reference systems, S and S'.
S' move relatively to the right with constant speed
$v > 0$.
An event arises on the common x-axis, x'-axis.
I usually consider an event as a short light signal.

I hope you agree with me when I say that :
An event (in this context) is a physical phenomenon that is independent of the two reference systems, S, S'.

The event can occur anywhere on the x, x'-axis.
We have set before that in SC1 it emerged in S'-origo and in SC2 it emerged in S-origo.

My question to you and all other relativists are:
How can an event affect the course of the clock? (time

dilatation)
How can an event affect a physical distance? (length contraction)

My letter No. 17:
Ulf Gran:
I am sorry Jan but I can't spend more time trying to explain this elementary physique to you, but I may refer you to a good text book on the subject, for example Rindler.

Me:
Why should you believe in Rindler in the place for Randy Harris?
For I believe that the simplest models are most suitable for finding errors in a theory.
...
Professor, you try to distance yourself from the basic concepts within SR.
You see that none of us give up. It is good!
But if we have opposite views on one and the same physical phenomenon, about one and the same mathematical concepts, then one of us is wrong.
Then maybe a judge would be needed. That is why I strive for a direct gathering.

I now want to present one of my evidence when it comes to LT, then the discussion about LT ends.

We start from two inertial reference systems, S and S'. S' move relatively S with constant speed $v > 0$, to the right in our model.

Our model represents reality. It is the same reality and model used in most derivations of LT.
For this model, the classic mathematical transformation between two coordinate systems applies.

$x = (distance\ between\ S\ and\ S') + x'$

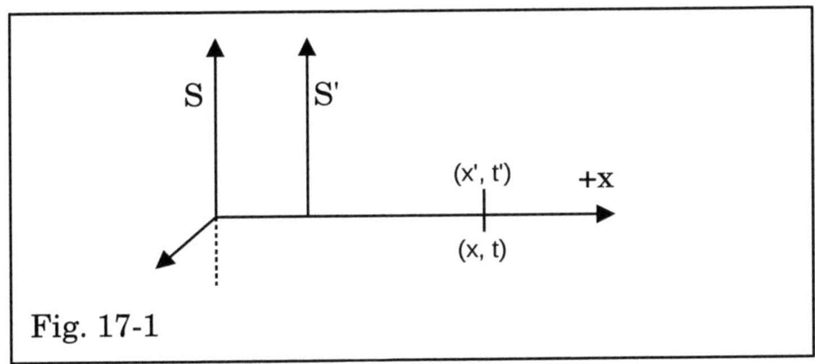

Fig. 17-1

So this transformation applies !!!

From LT results

distance between S and S' $= vt$ or

distance between S and S' $= vt'$

so we either have

$x = vt + x'$ or

$x = vt' + x'$

And if one now applied LT

LTx': $x' = (x - vt)\gamma$

Matematics: $x' = x - vt$

$\rightarrow \gamma = 1 \rightarrow v = 0.$

LTx: $x = (x' + vt')\gamma$

Matematics: $x = x' + vt'$

$\rightarrow \gamma = 1 \rightarrow v = 0.$

How do you explain this?

I consider and have proven it in many different parts of my research that LT is a creation that does not verify our reality, they cannot be applied to our reality!

My letter No. 18:
Ulf Gran:
I don't know if I should laugh or cry when you writing:
...
Me:
I think you don't understand my question.
It doesn't matter if the speed is small or large or zero.
In every moment there is the relationship

$x = (distance\ between\ S\ and\ S') + x'$

It's simple math.
Ask some students.
I don't want to tell you that you should take a basic course in mathematics.

(distance between two points on the x-axis in a Cartesian reference system)

Nils Andersson
n.a.andersson@soton.ac.uk
University of Southampton

My letter No. 1: 2021-12-18
Dear Professor Nils Andersson!
I write to you in Swedish and I write about your article in Kosmos 2021, Swedish Physician's yearbook.

Your article is titled *How do we know that Einstein was right?*
In your article you go through some historical facts, but you also touch on resistance to the theory of relativity.
This is dealt with in the article by Anna Davour *Why do so many people think that Einstein was wrong?*

I am one of them and my research concerns **the special theory of relativity, SR**.
On the University of Southampton website it says that you are the Head of General Relativity Group. But I hope you are familiar with SR as well.

Everything new that shows up about SR I read and

analyze based on my knowledge in mathematics and physics. Mathematics related to SR is simple. Physics as well.
So I read your article and here are my comments:

I start with my motto:
When studying physical phenomena, a mathematical model of these phenomena is created. In such a model, built-in physical laws are held together by mathematical tools.
If the description of the physical phenomenon is correct, the mathematical model is without contradictions and paradoxes.

Page 14:
"... if we measure a length range on a yardstick in motion ..."
My question: How do we measure it? I would like to see a picture of it.

Page 15:
Here you have a picture on three points and a light signal that is based on the origin of the reference system. You write:
"Since the light moves at the same speed in all directions, the signal when it arrives at point A has also reached points B and C."
My point of view: This can only happen if the reference

system is in absolute rest!

Page 19:
"If a beam of light comes in through a small window on the side of the rocket, the path of light will be bent against the floor"

It would fit well with a picture here, a model.
...
"... how the curved room and the elastic time ..."
Whose time is meant here? It is believed that in the theory of relativity, time is relative.

Page 27:
Here you talk about using SR within GPS. I have read about this argument for SR in other contexts as well. My question is: How can SR be used here when all satellites move in circular paths and SR applies to reference systems that move at a constant speed on a line (inertial reference system)?

Page 34:
“Was Einstein right? Obviously he had. "
...
"At the same time, it is clear that Einstein was wrong"

My letter No. 2: 2021-12-18
Me:

You write:
Thanks for the letter. As you probably understand, we have different perceptions in terms of relativity.

If we were to have the same view I would not have written to you.

As I see it, it is a theory that has been tested in many different ways and has clear support from the experimental side. You are certainly aware of that.

Please note that I only talk about the special theory of relativity, SR. Can you refer to an experiment that verifies SR?

But the ideas may seem paradoxical.

In physics and especially in mathematics there should not be paradoxes. See my motto from the last letter. Not even the ideas should seem paradoxical. If in a theory there is paradoxical elements then you have done something wrong, you have thought wrong, you have created a mathematical model incorrectly.

After having taught the subject for more than 20 years, I am well aware of the different pits you can get stuck in.

I have great respect for researchers, I respect their knowledge.
I am always overwhelmed when I read about new evidence, new inventions they present.
...
My analysis of *Introducing Einstein's Relativity* by Ray D'Inverno.
See chapter *Analysis of K-Calculus*... in this book.
...
Best professor, thanks again.
I am waiting for your arguments and I hope you have some time and patience to answer me.

It was long time I wanted to talk/exchange with a mathematician who also engages with SR.
For SR is a mathematical building in the first place, especially LT.

My letter No. 3: 2022-01-14
I have been waiting and waiting for your answer. I hope you have some extra time and explains, argues, regarding parts of my article that you looked at and which you think is wrong.

My letter No. 7: 2022-02-14
I have been waiting and waiting for your answer. It's been a month since I sent my analysis of the second chapter from the book *Introducing Einstein's relativity*;

Ray d'inverno, 1992.

I hope you had some time to read my analysis, judge my arguments.
So I'm waiting for your comments.

Frank Close
frank.close@physics.ox.ac.uk
University of Oxford, UK

Nils Andersson
n.a.andersson@soton.ac.uk
University of Southampton, UK

Ulf Gran
ulf.gran@chalmers.se
University of Gothenburg, Sweden

My letter No. 1: 2022-06-08
Dear Professors
I'm writing a joint letter to the three of you.
I have written before to each of you individually.

I wrote to **Frank Close** for the first time about two years ago.
Quotes from his first answer:
"I have received countless messages over decades with

claims that special relativity is wrong. I uniformly say the following: it is experiment that decides. If your theory has some experimental consequence that distinguishes it from relativity then let experiment decide. Otherwise any argument is sterile."

I did not get any more answers from him.
But I have been inspired by his book
Nothing: A Very Short Introduction.

I wrote to **Nils Andersson** for the first time in December 2021 regarding his article in Kosmos 2021, the Swedish Physicians' Society's yearbook.
"Since the light travels at the same speed in all directions, the signal when it has reached point A has also reached points B and C."
My point: This can only happen if the reference system is in absolute rest!

After a few letters, I received tips from Nils Andersson about the book *Introducing Einstein's Relativity* by Ray d'Inverno. I have analyzed chapter 2 and written my views and pointed out where somewhere in the book in the k-caculus error occurs.
I have sent my article about this to both Nils Andersson and Frank Close.
I have not yet received any answers with objections!

I wrote to **Ulf Gran** for the first time in early 2019. We have sent just over 20 letters to each other. But I have not received any answers to my specific questions.

That you teach / have taught the special theory of relativity and that you defend it feels very regrettable precisely because you do not answer to arguments that show that this theory is incorrect, that it is nonsense.

I expect you to judge my arguments against the special theory of relativity and to treat them objectively and with respect in the same way as I write to you.

With respect to science!

My letter No. 2: 2022-07-12
Dear Professors

It's been a month since I sent my letter to the three of you.
I have not received any response from any of you.
I understand that it's summer now, that you might have a holiday. I do not want to disturb you in your private life.

In response to my letter from 2019-03-26, Ulf Gran writes the following:

My tip is that you take a university course in SR, after that I think you got an answer as to why there are no problems with SR, and that SR is definitely not nonsense.

I have completed a course in the special theory of relativity and received my certificate which I attach. The course can be found at the link https://worldscienceu.com/courses/special-relativity-world-science-u/

The consequence: I am even more convinced that the special theory of relativity is nonsense! That's how it is!

But convince me three of you by answering my questions.

Question 1: I asked it to Ulf Gran in my letter to him from 2022-04-27.
We consider the following thought experiments:
- two inertial reference systems are located on the common x, x'-axis at a distance d from each other.
- an event occurs on the x-axis at a distance D from the S-origin.
See picture SCDd:

We have the following relationships:

$x = D = x' + d$

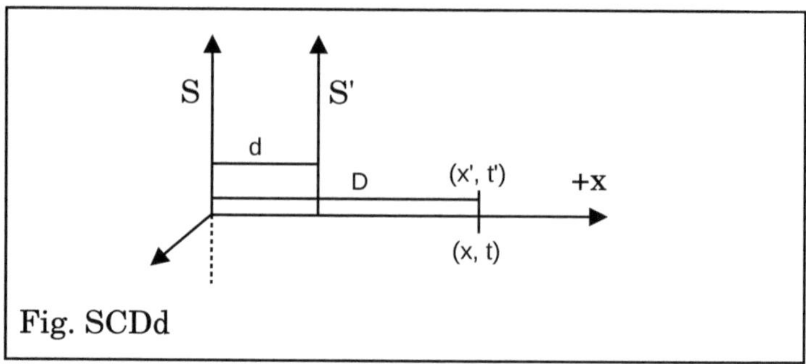

Fig. SCDd

But what is our definition for the time coordinates, t and t' ?
S and S' are at absolute rest relative to the x, x'-axis, $v = 0$.
What moment should we consider as $t = 0,\ t' = 0$?
It is not self-evident and whatever definition and convention we adopt for the time coordinates, we can get so many different answers in our further derivation.

For me, it feels most logical to use the t-coordinate as the time when S gets information that the event at point x occurred, when the light signal from this event reaches S. In similar cases, I define the t'-coordinate, as the time when S' gets information that the event at the point x, x' occurred when the light signal from this event reaches S'.

But what do we do if S and S' move relative to each other at speed $v > 0$?
Imagine the corresponding figure as Fig. SCDv where we replace d with $vt = vt'$.

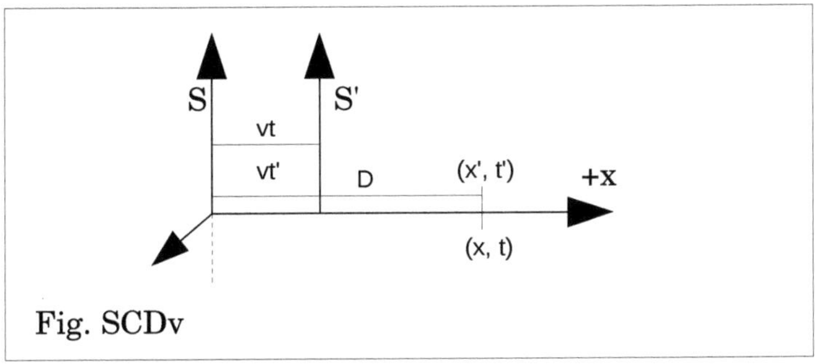

Fig. SCDv

Mathematically, the two figures are identical.
Please give me a definition of time coordinates.
Thanks.

I am waiting for the answer from Nils Andersson and Frank Close regarding my analysis of k-calculus from the book
Introducing Einstein's Relativity by Ray d'Inverno.

With respect to science!

15. References

[15.1] *Special Relativity in Modern Physics*, chap. 2, Harris R 2008

[15.2] *Special Relativity is Nonsense*; third edition; Jan Slowak; 2020

[15.3] *Introducing Einstein's Relativity;* Ray d'Inverno, Chapter 2; 1992

Jan Slowak: That is why theory of special relativity is nonsense!

16. My cetificate in SR

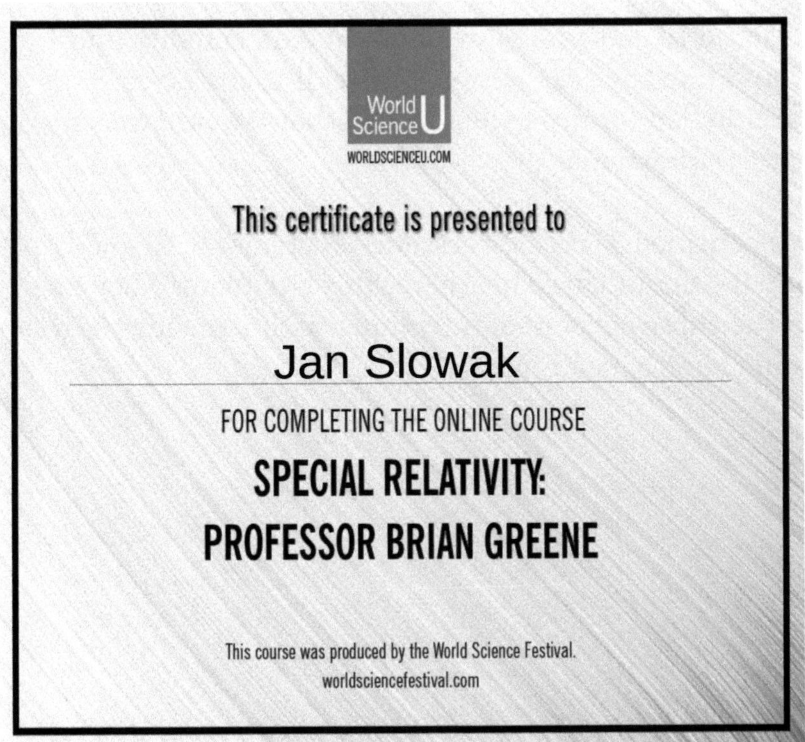

17. A small question for great researchers

I intend to do some more thought experiments and I would like the researchers who call themselves established (or are called so by others) to answer my 3 (three) questions.

We consider 3 (three) reference systems, S, S' and S''. Say it's about three brothers, three triplets. They are in the three corners of a triangle with two sides of the same length, see Fig. 17.1.

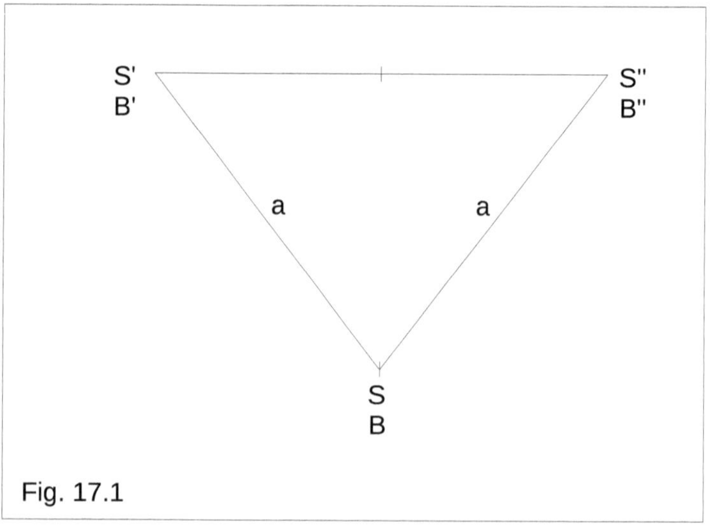

Fig. 17.1

The points S, S' and S'' are at absolute rest relative to each other and at absolute rest in space.

We have distance $d(SS') = d(SS'') = a$.
When the experiment begins, their clocks are reset:

$t = t' = t'' = 0.$

We consider S as stationary, at absolute rest in space. The three thought experiments that follow always start from Fig. 17.1, they have the same input conditions.

TE1
First thought experiment:
B' travels towards S and back at constant speed v'. Just do the math and record how much B has aged during this time. It is the same as calculating how much time has passed in S.
Question Q1:
What value does t have when B' is back in S'?

TE2
Second thought experiment:
B'' travels towards S and back at constant speed v''. Just do the math and record how much B has aged during this time. It is the same as calculating how much time has passed in S. $v'' = 2v'$.
Question Q2:
What value does t have when B'' is back in S''?

TE3

Third thought experiment:

Both B' and B'' start against S with the same conditions as in TE1 and TE2.

When B'' reaches S, B' is halfway to S.
When B'' is back in S'', B' reaches the point S.

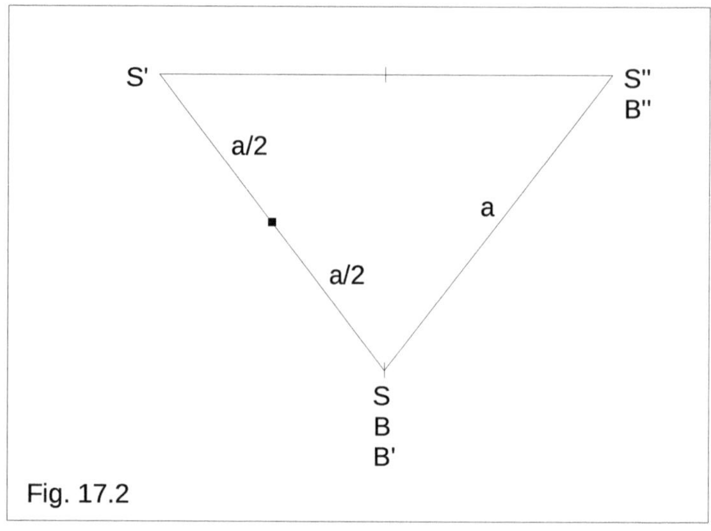

Fig. 17.2

Question to established researchers:

Question Q3:

What value does *t* have when B'' is back in S''?

How much does time *t* show relative to S' and how much does it show relative to S''?

Help: There is only one time in S!

Send your answer to jan.slowak@gmail.com